FARMERS' SEED PRODUCTION

Farmers' seed production

New approaches and practices

C.J.M. ALMEKINDERS and N.P. LOUWAARS

Practical
ACTION
PUBLISHING

INTERMEDIATE TECHNOLOGY PUBLICATIONS 1999

Practical Action Publishing Ltd
25 Albert Street, Rugby, CV21 2SD, Warwickshire, UK
www.practicalactionpublishing.com

© Intermediate Technology Publications 1999

First published 1999\Digitised 2013

ISBN 13 Paperback: 9781853394669
ISBN Library Ebook: 9781780442150
Book DOI: http://dx.doi.org/10.3362/9781780442150

Since 1974, Practical Action Publishing has published and disseminated
books and information in support of international development work
throughout the world. Practical Action Publishing is a trading name
of Practical Action Publishing Ltd (Company Reg. No. 1159018), the
wholly owned publishing company of Practical Action. Practical Action
Publishing trades only in support of its parent charity objectives and any
profits are covenanted back to Practical Action (Charity Reg. No. 247257,
Group VAT Registration No. 880 9924 76).

Contents

Foreword   xi

Contributing authors   xiii

INTRODUCTION   xv
    Why write a book on small-scale seed production?   xv
    The objectives of the book   xvi
    Who is the book for?   xviii
    Outline of the book   xviii

**Part A: Local seed systems: description, analysis and
experience**   1

1. The development of seed systems   3
  1.1 The history of seed selection in agriculture   3
     1.1.1 The development of crops and varieties   3
     1.1.2 Self-fertilizing, cross-fertilizing and vegetatively
       propagated crops   6
     1.1.3 Centres of crop origin and diversity   10
     1.1.4 Local seed systems   14
     1.1.5 Local varieties for multiple uses and preferences 14
     1.1.6 Local varieties for variable environments   15
  1.2 The development of formal seed supply systems   18
     1.2.1 Industrialized countries   18
     1.2.2 Developing countries   22
     1.2.3 The use of local and modern varieties in local
       systems   25
     1.2.4 Linking the formal and the informal sectors   28

2. Improving local seed systems   33
  2.1 The need to strengthen the local seed system   33
  2.2 Crop improvement: participatory plant breeding and
  variety selection   35
     2.2.1 Introduction   35
     2.2.2 The introduction of new materials   35
     2.2.3 Improving and maintaining varieties   36
     2.2.4 Participatory crop improvement   38
     2.2.5 Examples of PVS and PPB   42
  2.3 Improving seed production   46
     2.3.1 On-farm production and storage of seed   46

	2.3.2	Supporting specialization in seed production	48
2.4	Variety diffusion and seed exchange		50
	2.4.1	Farmers' existing seed sources	50
	2.4.2	The strategy for strengthening traditional seed exchange	52
	2.4.3	Demonstration plots	54
	2.4.4	Seed kits	54
	2.4.5	Seed fairs	55
	2.4.6	Seed production with key producers	56
2.5	Support to local seed security		56
	2.5.1	Seed banking	56
	2.5.2	Seed from relief programmes	58

3. Biodiversity and *in situ* conservation 61
   3.1 Biodiversity 61
   3.2 Biodiversity in agriculture 62
   3.3 The loss of biodiversity 63
   3.4 Conservation strategies 65
       3.4.1 *Ex situ* conservation 65
       3.4.2 *In situ* conservation 67
       3.4.3 On-farm conservation 68

4. Support from the formal sector 72
   4.1 Contributions by the formal seed sector 72
   4.2 Legislation 73
       4.2.1 Different regulations 73
       4.2.2 Seed laws 73
       4.2.3 Possible effects of seed laws on local seed initiatives 76
       4.2.4 Phytosanitary laws 77
       4.2.5 Plant Variety Protection: plant breeders' rights 77
       4.2.6 Related issues 79
   4.3 Contributions by the formal farmer support services 80
       4.3.1 Extension 81
       4.3.2 Credit 81
       4.3.3 Inputs 82
   4.4 Contributions to local seed systems at the policy level 83
       4.4.1 Economic policies 83
       4.4.2 Research policy 83
       4.4.3 Seed policy 84

Part B: Technical aspects of seed production 85

5. Seed quality 87
   5.1 The significance of seed quality 87

	5.2	Aspects of quality	87
		5.2.1 Physiological seed quality	88
		5.2.2 Sanitary seed quality	90
		5.2.3 Analytical seed quality	90
		5.2.4 Genetic seed quality	91
	5.3	Monitoring seed quality	93
6.		Seed production agronomy	94
	6.1	General principles	94
		6.1.1 The similarities and differences between crop and seed production	94
	6.2	Land preparation and crop establishment	95
		6.2.1 The choice of field and isolation	96
		6.2.2 Environmental conditions	97
		6.2.3 Tillage	97
		6.2.4 The choice of seed	97
		6.2.5 The timing of planting	100
		6.2.6 Plant population	100
	6.3	Crop cultivation and protection	101
		6.3.1 Fertilizers	101
		6.3.2 Pest, disease and weed management	103
	6.4	Special groups of crops	108
		6.4.1 Biennials	108
		6.4.2 Pasture species	109
		6.4.3 Perennial crops	109
7.		Harvesting, processing and storage	112
	7.1	Harvesting	112
	7.2	Drying	112
	7.3	Storage	113
	7.4	Seed cleaning	115
	7.5	Chemical seed treatment	116
8.		Seed and variety selection	119
	8.1	Options for selection	119
	8.2	Variety maintenance	121
		8.2.1 The maintenance of diversity	121
		8.2.2 The maintenance of uniformity	124
		8.2.3 Self-fertilizing crops	124
		8.2.4 Vegetatively propagated crops	126
		8.2.5 Cross-fertilizing crops	126
		8.2.6 Semi-cross-fertilizing crops	131
	8.3	Crop improvement	131
		8.3.1 General	131
		8.3.2 Participation from the start	132

8.3.3   Introducing participation in an existing
programme   133

9.   Technical aspects of seed enterprise development   135
    9.1   Introduction   135
    9.2   The demand for seed   136
    9.3   Costs and price/quality ratio   138
    9.4   Competition   141
    9.5   Distribution   143
    9.6   Promotion   143
    9.7   Seed production planning   144

10.   Seed quality testing   147
    10.1   Introduction   147
    10.2   Sampling   148
    10.3   Moisture content   149
    10.4   Seed purity   151
    10.5   Germination capacity   154
    10.6   Seed vigour   156
    10.7   Seed health   157
    10.8   Genetic composition   158

Part C: Working with farmers   161

11.   Problem diagnosis   163
    11.1   Introduction   163
    11.2   Orientation   163
        11.2.1   General orientation   163
        11.2.2   Further orientation – an informal survey   166
    11.3   Planning and implementing action   167
    11.4   Survey of a local seed system   170
    11.5   Designing seed system surveys   171
    11.6   Preparation of the questionnaire and data collection
sheet   180
    11.7   Gender analysis   184
    11.8   Presentation of results from a survey   188

12.   Participatory experimentation   189
    12.1   Priority setting and planning   189
        12.1.1   Diagnosis   189
        12.1.2   Planning the experiments   189
    12.2   Experimentation with varieties: variety evaluation   191
        12.2.1   Introduction   191
        12.2.2   Different types of variety trial   192
        12.2.3   Which materials to include   197

	12.2.4	Participatory evaluation methodologies and tools	198
	12.2.5	The need for statistical analysis	199
	12.2.6	Field layout and design	200
	12.2.7	Examples of planting designs	202
12.3	Case studies of variety improvement through selection		202
	12.3.1	How to use the presented examples	202
	12.3.2	Case 1: Improvement of degenerated varieties	203
	12.3.3	Case 2: Generating uniformity for the market	204
	12.3.4	Case 3: Responding to changing environmental conditions	205
	12.3.5	Case 4: Increasing yields in existing cropping systems	208
12.4	Experimentation with seed production		209
	12.4.1	Case 5: Improving farm-produced bean seed in Costa Rica	210
	12.4.2	Case 6: Improving on-farm seed production and the use of millets	211
	12.4.3	Case 7: Building on traditional seed flows	211
12.5	Experimentation with seed quality measurements		212
	12.5.1	Standardization	212
	12.5.2	Seed testing to evaluate observed problems	213
	12.5.3	Seed testing as part of an experiment	213
	12.5.4	Seed testing as an evaluation tool	214
	12.5.5	Testing test methods	214

Part D: Crop-specific options 215

Barley	217
Bean	219
Cassava	222
Chickpea	223
Cowpea	224
Groundnut	226
Finger millet	229
Maize	230
Rice	234
Pearl millet	237
Potato	239
Sesame	242
Sorghum	244
Soybean	247
Sweet potato	249
Sunflower	251
Wheat	252

VEGETABLES
Alliums   256
Cruciferous vegetables   259
Cucurbits   261
Solanaceous vegetables   263
Local leafy vegetables   266

Part E: Appendixes   271

Appendix 1.   GLOSSARY and ABBREVIATIONS   273
Appendix 2.   CROP NAMES   280
Appendix 3.   REFERENCES   281
Appendix 4.   RECOMMENDED READING   284
Appendix 5.   ADDRESSES   289

Foreword

Seed is a crucial input in agriculture. It can grow into an entire plant with roots, stems and leaves, and produce seed again: it is a vehicle for genetic information and it is reproducible. The genes in seed determine how the crop will develop, how it will react to fertilizer, pests, diseases and droughts. They determine its cooking time and taste, as well as its storability.

It has become apparent that many seed projects designed during the last few decades have not been successful. It is increasingly recognized that, although having significant shortcomings, on-farm seed production is a valuable alternative. It is often the best option for most of the farmers in Africa, Asia and Latin America. This recognition means that we must reconsider approaches and practices for developing seed production, and in particular recast the tasks of the different actors in the process. Although several studies and publications on new approaches to co-operation in development have appeared, the translation of these to seed-technology practices is lacking.

Taking the local seed system as a starting-point offers many opportunities for improving seed supply. In a majority of the situations where farmers rely on their own or locally produced seed, limitations exist in seed quality or access to good seed. These can be addressed by making use of local knowledge and resources. Seed technology, when properly applied, can provide a valuable additional input, offering a range of alternatives which can be tested and incorporated into small-scale farmers' seed production when successful.

Improving farmers' seed production implies working closely with farmers. That demands an additional input to feed technology: participatory approaches that combine 'local' knowledge and 'scientific' seed technology in order to develop improved seed production which is adapted to the local situation. This book intends to introduce both these aspects: the local perspective and scientific technology. It is principally meant for people who work directly with farmers in improving their seed supply. The book relates concepts on local seed systems to ideas for action, and presents the basics of seed quality, seed production, storage and selection needed for improvement. The combination of these provides suggestions for action in which formal seed technology may be used to meet the needs of local seed supply. These suggestions are by no means a guarantee of success. Experience in this field is still limited, and every situation is different and requires a creative process of analysis, prioritizing limitations, suggesting solutions, experimentation and diffusion of results. It is not the intention of the authors to present a blueprint approach to improving farmers' seed production. The suggestions are made with the hope that they may generate more experience, to be shared with others in some way.

This book sprang from the need for basic information which was encountered in the field by the Small-Scale Seed Project in the SADC region. The experiences and needs of this project have motivated and supported the authors of this book to combine experience in participatory approaches and seed-system development, together with seed technical information in one volume. However, to have the idea that an interesting topic should be written about is only the beginning of a great number of things to consider. One of them, ranking prominent amongst others, is to secure the financial support to carry out the task. It is therefore gratefully acknowledged that funding on GTZ/BMZ, CPRO and NEDA permitted the publishing of this book.

It is hoped this book will improve understanding of the values of both farmers' and formal seed systems, in order to stimulate interaction between them. The fruits of these interactions will contribute both to an increased sustainability of farming systems and to the advancement of the formal seed sector and national policies guiding agricultural development.

Ortwin Neuendorf
GTZ – Small Scale Seed Project in SADC, Harare

Contributing authors

Conny J.M. Almekinders is an agro-ecologist who is presently working at the CPRO-DLO Centre for Genetic Resources, The Netherlands. She is involved in the development of activities related to the *in situ* conservation of agro-biodiversity, such as participatory breeding, local seed supply. She is the principal and coordinating author of Parts A and C of this volume and contributed to other sections of the book.
Address: CPRO-DLO, P.O. Box 16, 6700 AA Wageningen, The Netherlands

Walter S. de Boef was trained as a breeder and rural sociologist. He has worked at the CPRO-DLO Centre for Genetic Resources, and was involved in the development of *in situ* conservation of agrobiodiversity, participatory plant breeding and local seed supply systems. He presently works at the Royal Tropical Institute. His principal contribution to this book was the section on biodiversity and *in situ* conservation.
Address: KIT, Mauritskade 63, Amsterdam, The Netherlands

W. Joost van der Burg is a seed technologist of the Department of Reproduction Technology of CPRO-DLO. He has a long-standing experience in seed quality testing, both in developed and developing countries. His principal contribution to this book was the chapter on seed quality testing (Chapter 10). He also contributed to sections on rice and sorghum in Part D.
Address: CPRO-DLO, P.O. Box 16, 6700 AA Wageningen, The Netherlands

Elizabeth Cromwell is an agricultural economist and a Research Fellow at the Overseas Development Institute, London, an independent non-governmental think-tank on development policy. Her particular interests lie in assessing policy for local-level crop genetic resources conservation and seed supply. Her principal contributions are to Section A of this book.
Address: ODI, Portland House, Stag Place, London SW1E 5DP, United Kingdom

Lisa Leimar-Price is a cultural anthropologist with expertise in gender, agricultural systems ethnoecology and biodiversity. She carried out extensive research in South-east Asia on women and management of biodiversity. Formerly a senior scientist at IRRI, she is now a senior lecturer in Wageningen.
Address: Dept. Gender Studies, Chairgroup Sociology, Wageningen Agricultural University, Hollandseweg 1, 6706 KN Wageningen, The Netherlands

Niels P. Louwaars is a breeder with field experience in seed production programmes in developing countries. He currently focusses on seed policies and the regulatory and organizational issues of seed supply, integrating the strengths of local and formal knowledge and materials. He is the principal and co-ordinating author of Parts B and D of this book and contributed to other sections of the book.
Address: CPRO-DLO, P.O. Box 16, 6700 AA Wageningen, The Netherlands

Robert Tripp is an anthropologist and a research fellow at the Overseas Development Institute. He works on agricultural research and extension issues and has recently been involved in studies of seed policy and seed regulation.
Address: ODI, Portland House, Stag Place, London SW1E 5DP, United Kingdom

Responsibility for the final text remains with the principal author-editors.

ACKNOWLEDGEMENTS. The author-coordinating editors of this book have used more in this book than the sole texts of the above-mentioned contributing authors. Over the past several years, the co-ordinators have had the opportunity and pleasure of interacting intensively with most of the colleague-authors and with many of the authors of selected texts, examples and cases used in the publication. This exchange of experiences and insights has importantly stimulated and enriched their thinking; the ideas and approaches described in this book are the products of a shared mind-development. The editors thank all the authors who contributed original texts, and those published elsewhere.

The editors wish to thank Mr Ortwin Neuendorf, GTZ, project manager of the Small-scale Seed Project in SADC, for his support in the generation of this publication. They also thank Edward Lulu, chief seeds officer, Seed Control and Certification Institute, Zambia, and Michael Turner, head of the Seed Unit at ICARDA, for their comments on a draft of the manuscript.

Introduction

Why write a book on small-scale seed production?

Seed: the most crucial input

Seed can produce an entire plant with roots, stems and leaves, and produce seed again: it is a vehicle for genetic information and it is reproducible. A handful of seed offers the potential to grow a crop this year and for years to come. Improved genetic information written in the genes of the seed can result in better yields this year, but also in the following years. On the other hand, contamination by pathogens, suboptimal production or storage practices, or genetic degeneration through unsuitable seed selection practices can negatively affect yields over more than one season.

This capacity to reproduce itself and carry over effects to following seasons is seed's unique, and crucial input in agricultural production. It also makes it an attractive tool for agricultural development. Seed projects and programmes have been and are on the agenda of most agricultural development organizations. Despite this importance, there are relatively few books which tell you 'how to produce seed'. Most of the seed publications to date discuss the organization of specialized, large-scale formal seed systems, and implement advanced seed technology for specialized seed production.

This book focuses on seed production by small-scale farmers. After all, most seed in this world is produced by farmers themselves. The major principle of seed production is simple for most situations: good crop husbandry produces good seed. If seed production is so simple, what makes seed production by small-scale farmers so different, and why is this book necessary?

The complexity of farmers' seed production

With 'farmers' seed production' we essentially refer to growing a crop, part of which part is saved as seed for own use, and its harvesting, storage and seed selection. These activities take place on-farm and are usually realized by members of the household. Exchange of seed with others is mostly confined to family and local contacts, although sometimes large distances are travelled to secure seeds. Because these seed activities are fully integrated with other activities in the farming system and its environment, issues of on-farm seed production should be placed in a wider context. These seed issues and their context are referred to in this book as *farmers' seed systems*, *local seed systems*, or *informal seed systems*.

Because of the context, farmers' seed production is more complex than suggested by the phrase that good husbandry produces good seed. It involves a wide range of activities and decisions, and it takes place on farms where multiple objectives have to be met. Needs for home consumption, fodder, construction materials and markets have to be balanced for a number of crops, animals and off-farm activities. Decisions about seed production and seed-saving depend on the possibility of keeping part of the harvest from being consumed, and on off-farm sources of seed of acceptable quality. The options for giving special attention to seed production are limited where farmers have to deal with a multitude of other issues. The technology appropriate for producing good-quality, farm-saved seed is therefore not the same as in specialized seed programmes. This means that for improving farmers' seed production, a thorough understanding of the whole farming system and its agro-ecological and social environment is needed.

Recognition of the importance of on-farm seed production, increasing awareness of local knowledge and practices as a resource for development, and growing evidence of the importance of farmers' participation provide a solid basis for agricultural development, for the adaptation of concepts, and for developing practices and tools in improving seed supply for small-scale farmers in developing countries. This has resulted in changing objectives and approaches: taking the local seed systems as a starting-point and aiming at improving these systems, making use of farmers' capacities and knowledge, and selecting and adapting from the formal system what is suitable.

The changes in objectives mean that the attention is shifted for different aspects. For the technology of seed production, this book draws on technologies from the formal, conventional seed sector, focusing on the aspects that are particularly important for on-farm seed production. The genetic quality of seed, an aspect which is taken care of by breeders in the formal system, but which is also of concern to farmers in local systems, is receiving ample attention. The discussion of participatory breeding and the practices related to it are essential to this focus.

The objectives of the book

Taking the local seed system as a starting-point
With this book we attempt to generate an understanding of what local seed systems are. The aim is to make clear that local seed systems have advantages for the farmer, but may have serious shortcomings, providing possibilities for and requiring improvements. Such improvements should aim at a better quality of seed and a higher seed security. The target may not always be seed production by individual farmers, but also the diffusion of

the seed, the variety and the technology to other farmers. By implementing these ideas in collaboration with farmers, using participatory methodologies, both farmers and technicians will learn and increase their capacity to develop new technologies.

New approaches, no recipes, but a challenge to the user

This book cannot and does not intend to present recipes for improving farmers' seed production of particular crops. It aims to contribute to a better understanding of what farmers' seed production systems are about, and what their strengths and weaknesses may be. The challenge for the user of this book is to share with farmers the understanding obtained from reading this book, to study the existing seed system and identify limitations and areas for improvement. Some of the suggestions for action may be tried out, evaluated and modified to fit the reality of a particular situation. The fact that each situation, each crop, each environment and socioeconomic condition creates a particular, unique situation is another reason why a blueprint or recipe approach will never work.

Linking different systems and actors

Another principal objective of this publication is to stimulate the co-operation of farmers and farmers' organizations with scientists and technicians in the formal sector.

The strengths and potential of the local seed system and of farmers' capacity to produce their own seeds are not fully recognized in formal circles. National seed programmes, however, are not very effective in supplying seed directly to most small-scale farmers.

Linking formal and farmers' seed systems offers challenging opportunities to overcome the weaknesses and optimizing the strengths of both systems. Developing suitable linkages with the formal sector may offer the local system access to a basket of options in the form of materials and technology not otherwise available and useful for improving the sustainability of local systems.

The option of linking with the local seed system to provide a range of seeds and varieties to small-scale farmers, and relying on the strength of these systems for further diffusion, may result in an approach which makes the formal seed sector more effective. Often, however, differences in approach lead to an inability to co-operate. Seed specialists in the formal sector may stick with certification systems and may not recognize farmer-specialists and rural development workers as their equals. On the other hand, there is a trend to romanticize small-scale farming, farmers' knowledge systems and local germplasm. Both attitudes lead to missed opportunities that co-operation would offer. Scientific knowledge about seed production, agronomy, seed physiology and seed pathology can supplement existing farmers' knowledge and materials. On the other hand, farmers' knowledge can enrich the scientific agricultural research system. This is particularly true in variety evalutions, but also in local seed-storage technology. Looking for complementarity and integration of local and formal seed systems is therefore likely to be beneficial for both.

Who is the book for?

The book is aimed at farmers and farmers' groups who may want to improve their seed system. The book is, however, especially written for organizations and persons that are in the position to link farmers with outside knowledge and technology, such as extension agents, national and international NGO personnel, and farmer cooperative workers. They may have the social and technical resources to analyse limitations, suggest improvements and test adapted technologies in a participatory manner. For those this book may present valuable ideas and suggestions for action, but as mentioned earlier, it does not provide a blueprint for such action. Policy makers may find the book valuable as an introduction to actual discussions related to seed supply and small farmers in developing countries, an issue which they may want to address and integrate into national seed policies.

Outline of the book

This book consists of four parts.

Part A. The description of local seed systems is divided into three sections: the various aspects of varieties and seeds used by farmers, seed production and handling practices, and the mechanisms of seed exchange and diffusion. The definition of the sections is somewhat artificial since these different aspects are strongly related.

The relation of local seed systems with *in situ* maintenance of genetic diversity, and the international debate on access to genetic resources are discussed in separate chapters in Part A. These issues do not have a direct relation with the production and storage of the seed, but form the background information that anybody who is active in local seed supply should be aware of.

Part B. The second part of the book deals with the technical issues of seed production, handling, storage and selection to improve seed quality. Most of this part is based on conventional seed technology, attempting to apply this to conditions of farmers' seed production. As mentioned earlier, genetic aspects of seed production get ample attention, as these are an important aspect in on-farm seed reproduction and crop development. The approach of participatory breeding is translated into suggestions for practical experimentation.

Part C. The third part of the book contains practical guidelines on how local seed systems can be studied, analysed and improved. Suggestions and examples for surveys, experimentation on variety selection and improvement, seed production and seed quality are outlined.

Part D. The fourth part of the book contains the principal crop technical information, with special attention to those aspects which are relevant for seed production. It serves as a quick reference which should be read in conjunction with the former sections.

The References and Recommended Reading sections contain the key literature on local seed systems, participatory approaches and seed technology. These references are also the principal literature used in the preparation of this publication. As much as possible we have limited the references to the information that is easiest to access. We have not attempted to be exhaustive in covering the literature on the topics, but the given references should provide an overview of the most relevant accessible literature.

Conny Almekinders, Niels Louwaars

PART A

LOCAL SEED SYSTEMS

Description, Analysis and Experience

1. The development of seed systems

Two different type of seed supply systems, the local, informal system and the formal system exist today as a result of diverging developments in agriculture. The historic development of agriculture and agricultural crops explains this situation. Complementarities in the two systems indicate a need for linkage and integration to optimize the functioning of both.

1.1 The history of seed selection in agriculture

1.1.1 The development of crops and varieties

Earliest cultivation
Initially, man collected leaves, roots, tubers, fruits and seed, to be consumed with hunted meat and fish. The question why man started deliberately to plant seeds for cultivation will probably always remain open to speculation. Theories suggest that increasing population pressure was the principal reason, possibly triggered by changes in climate. Agriculture was probably first practised in the Middle East approximately 10 000 years ago. The general assumption is that it spread from there into Europe, Africa and Asia. However, archaeological findings indicate that agriculture may have begun independently in different parts of the world.

The first 'cultivated' plants may have grown from spontaneously germinating seeds and sprouting roots or tubers in waste heaps around homesteads. At a certain point in history grains or tubers were saved, with the purpose of sowing them at a determined place near the homestead. Saving the best seeds, roots or tubers from consumption, storing and planting them became associated with 'cultivation'. Practices of nurturing the crop by tilling the soil, weeding and watering, and adding manure must have developed as an understanding of plant growth and environment increased.

Natural selection
Initially only wild plants existed; crops and crop varieties as they are known today are a result of a combination of natural and farmer selection. The genetic characteristics of wild plants arise from the interaction between plants their environment. Natural selection is the process by which consistently more successful individuals are favoured from a population of different genotypes. The genotypes which are most successful in growth,

multiplication and survival produce most offspring and contribute more to the following generation. If this success is consistently genetically determined, the result is that, over time, the population will have more and more individuals which carry the genes for this success.

For example, in an arid environment, natural selection will favour the genotypes in a population which are most successful in reproducing under drought stress. If a particular disease prevails, then natural selection will favour the individuals which carry resistance genes.

Farmers' selection
Farmers can modify the natural selection process through the way they select the seeds or cultivate the crop. Early agriculturists must have selected and saved grains, tubers or roots that they liked most: the ones which were largest, tasted best and cooked fastest, seeds that did not easily shatter, from plants without thorns, which yielded highest, and showed the least insect and disease injury, etc. (see Figure 1.1). As a result, characteristics that determine the success of plants in surviving in a natural environment, like early shattering of seeds, disappeared from the cultivated plant populations. Bitter substances and thorns, making the plants unattractive for grazing animals, were eliminated from cultivated species such as lettuce. As agriculture developed, plants became genetically more adapted to cultivation. Weeding eliminated the slow and late emerging crop seeds, resulting in more rapid and even seed germination. Seeds with long dormancy would not germinate and did not contribute to the harvest in the next season.

For some characteristics farmers selected unconsciously. For example, the conscious selection for large ears in maize meant an unintentional selection for fewer ears per plant and less tillering. Farmers' selection often acts against natural selection pressure, as in the case of the seed shattering. For other characteristics, farmer seed selection supports natural selection, such as the selection of healthy looking seeds that may support the natural selection for some disease resistance. Thus, by repeated selection for the same characteristics over years and years, farmers developed plant populations that differed from those developed through natural selection pressure alone; seed selection was the key-activity in the process of crop domestication. Here 'seed' is used in a wide sense, also referring to roots and tubers used for planting.

A plant population that has developed over time as a result of natural and farmer selection practices is called a *farmers' variety, traditional variety, local variety* or *landrace* (see Box 1.1 and Section 1.1.4).

New genetic variation: mutations, introgression and hybridization
Without genetic variation a plant population cannot continue to evolve. New genetic variation is introduced into a population of crop plants through natural mutation, introgression from wild or weedy relatives,

4

Box 1.1 Definitions

Variety: a plant grouping which is distinct in one or more forms or functions from other such groups of a plant of the same species and which maintains these distinctions when reproduced; in this book synonymous with *cultivar* which is a cultivated variety.

Farmers' variety, traditional variety, local variety, or landrace: a population of plants resulting from the combination of environmental selection pressure (biotic and abiotic) and the cultivation and seed selection practices of the farmer over a period of time. A landrace is usually a complex heterogeneous population, but not necessarily so.

Improved variety: a term that is used in different ways by different people, and which is therefore not used in this book.

Modern variety: variety which is developed by trained breeders working through targeted generation of diversity, through crossing or other bio-technology tools, and selection.

High yielding variety: used as a synonym for modern variety, but actually an incorrect term since it implies that modern varieties always have high yields.

Seed: the generative or vegetative part of the plant used as propagation material. Correctly used, seed refers only to the plant part that developed from the flower after fertilization (true seed). This correct use is sometimes emphasized in this book by making separate reference to the seed and vegetative plant parts (roots, tubers, stems) used for propagation.

Propagation material: the overall term for any plant part used to grow a crop (seed, tuber, root or stem part).

Crop: in a narrow sense refering to cultivated and harvested plants. The forester may speak of a timber crop, the herdsman of a calf crop and the American Indians can speak of a wild-rice crop (Harlan, 1992).

Domestication: the process of genetic evolution that adapts the plant to cultivation and use. A fully domesticated plant or animal depends on man for survival.

Cultivation: caring for plants through any of the following activities: preparing the soil, planting, manuring, protecting, watering, followed by harvesting of the products the plant was cultivated for.

hybridization with other cultivars and the physical mixing of seeds from other varieties.

Mutations are naturally occurring very small alterations in the genes, which may or may not be visible. Mutations that are negative for the performance of the plant – as is the case in the majority of mutations – will be eliminated through natural or farmer selection. If mutations have a positive effect on the plant growth and development, the altered gene is likely to remain in the population and its frequency will increase. The mutation may even result in a different plant type, which the farmer selects and multiplies separately. Such positive mutations play a very important role in evolution.

When a crop plant is fertilized by pollen from a wild relative, genes may be introduced from the wild population into a crop variety. The reverse can also occur when the pollen of a crop plant fertilizes a wild relative. Such an introduction of genes from the wild or a weedy species, or vice versa, is called introgression. It is an important source of new genetic variation for crop plants where wild or weedy relatives occur in the borders around the field, which is the case for many crops in their centres of origin. Wild relatives of the maize, teosinte (*Zea diploperennis*) and *Trypsacum*, can still be found in some areas in Mexico and Central America, tomato can cross with wild relatives found in Central America and the Andes, and crosses between sorghum and wild relatives are reported for many parts of Africa.

Evolution
This generation of new genetic variation, in combination with natural and farmer selection, produces a continuous dynamic process of interaction between plant, farmer and environment. This shapes the local genepool and adapts it to the farmer's practices and preferences and to natural conditions. This process is called 'evolution'. Without genetic diversity in a genepool there is no evolution, so in order to maintain evolution in farmers' fields and the capacity of crops to adapt to changing conditions, sources of new genetic variation need to be maintained. This can only be achieved if the farming systems in which farmers produce and select their materials are also maintained (see Section 3.4.2 *in situ* conservation).

1.1.2 Self-fertilizing, cross-fertilizing and vegetatively propagated crops
Landraces of self-fertilizing crops (e.g. wheat, rice, beans) have developed from genetically diverse populations over time, into a mixture of lines, each line being a genetically homozygous genotype that produces genetically similar offspring (Box 1.2). Occasional mutation, cross-fertilization with pollen from another line (hybridization), or from a wild relative (introgression) can introduce a new gene. This fertilization will produce a genetically heterogeneous genotype; after a number of generations this heterogeneity will have disappeared and a successful new gene may have been integrated into a homozygous genotype (see Boxes 1.2 and 1.3).

In the case of cross-fertilising crops, the pollen will fertilize at random, by wind or insects. Unlike self-fertilizing crops, alleles will randomly recombine each season. As a consequence, genotypes will change, but the gene frequencies will remain stable if no particular natural selection or farmer selection intervenes (See Boxes 1.2 and 1.3).

(a) Triticum : different species (Source: Chrispeels and Sadava, 1994)

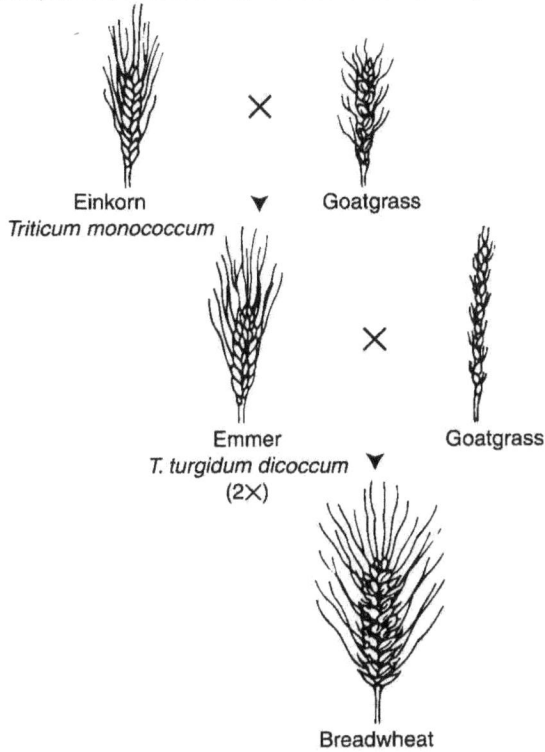

Einkorn   ▼   Goatgrass
Triticum monococcum

Emmer   Goatgrass
T. turgidum dicoccum ▼
(2×)

Breadwheat

(b) Brassica oleracea : different species (Source: Harlan, 1992)

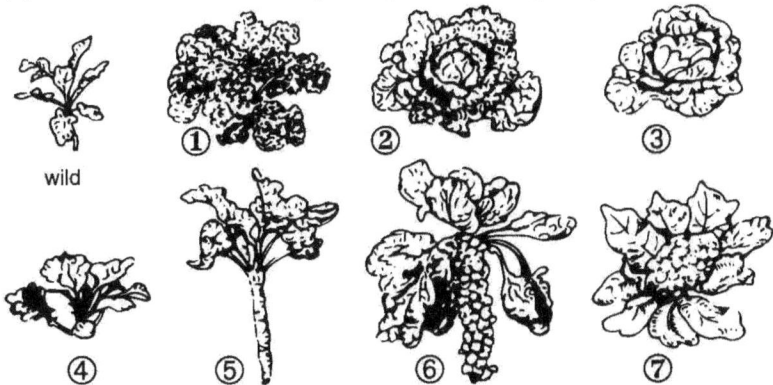

wild

① ② ③

④ ⑤ ⑥ ⑦

Figure 1.1 Examples of evolution. a. The development of breadwheat out of crosses between different species, as a result of natural and farmer selection. b. Wild and cultivated varieties of *Brassica oleracea* demonstrate the wide variation: farmers have selected plant types for the leaves (1), nicely formed heads (2, 3), stems (4, 5, 6) or flowers (7).

7

Box 1.2 Mendel and the pea experiments

Most of Mendel's work was done with garden peas (*Pisum sativum* L.). He used a number of varieties that differed from one another in many traits. The term *trait* refers to a character that occurs in two or more forms: for example the trait 'shape' occurring in the forms 'round' or 'wrinkled', or yellow or green seed colour, coloured or white seedcoat, inflated or constricted pod shape.

When Mendel crossed two varieties that differed in respect to one of the traits, he found that all the offspring resembled one parent and none looked like the other parent. For example, if he crossed plants growing from yellow seeds with plants grown from green seeds, all the offspring had yellow seeds. Mendel called this form of a trait *dominant*. The form that was not present in the offspring he called *recessive*.

The original parental plants are called the P (parental) generation. Their hybrid offspring are known as the F_1 (first filial) generation. When these F_1 plants were allowed to fertilize themselves or each other, they produce the F_2 (second filial) generation. When Mendel produced such F_2 generations, he found that both alternatives of each parental trait were present among the F_2 plants.

The recessive form that seemed to disappear in the F_1 generation was not eliminated, but reappeared in the F_2 generation. In each case, the ratios of the dominant and recessive form of a trait in the F_2 were close to 3:1. It was later explained that each plant has two copies of the gene and that these separate during gamete formation: each pollen grain or ovule has only one member of each pair. After fertilization, the gene of the pollen grain and the gene of the ovule together form a gene pair again. In the case of a self-fertilization the pollen and ovule cannot be of different plants.

Genes are located on filamentous structures called *chromosomes*. Humans have 23 pairs of chromosomes and maize 10. Fifty years ago it was found that genes are made of deoxyribonucleic acid (DNA) and that they contain the information to direct the synthesis of proteins. When the two copies of the genes are identical, the organism is said to be homozygous for the particular gene or trait. If they are different, they are heterozygous. These different forms of the gene are called *alleles* and the place of the gene on a chromosome is called its *locus*.

The symbols R and r (differing alleles of a same gene) are used to represent the contrasting alleles that cause the development of round and wrinkled seeds. If the two members of a gene pair differ, as they do in plants of the F_1 generation, half of the gametes will receive one allele and half the other. After fertilization, the ratio of the dominant and recessive individuals in the F_2 generation is 3:1. This ratio concerns the *appearance* of the seeds, i.e. their *phenotypes*; 3 round seeds: 1 wrinkled seed. If we consider their genes as shown by their genotypes, the ratio is: 1:2:1 (25% RR, 50% Rr, and 25% rr). RR and rr are refered to as *homozygous*.

By continuing self-fertilization the number of *heterozygous* individuals in a self-pollinator population declines (Figure 1.2a).

8

(a) Selfing after crossing of two homozygous parents

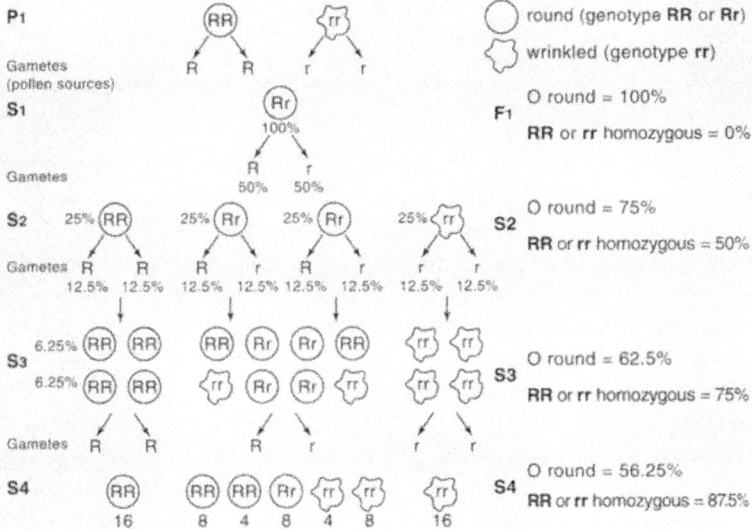

Phenotype:

P₁ — (RR) (rr)

○ round (genotype **RR** or **Rr**)

Gametes (pollen sources) — R R r r

⬡ wrinkled (genotype **rr**)

S₁ — (Rr) 100%

F₁

○ round = 100%

RR or **rr** homozygous = 0%

Gametes — R 50% r 50%

S₂ — 25% (RR) 25% (Rr) 25% (Rr) 25% (rr)

S₂

○ round = 75%

RR or **rr** homozygous = 50%

Gametes — R R | R r | R r | r r
12.5% 12.5% | 12.5% 12.5% | 12.5% 12.5% | 12.5% 12.5%

S₃ — 6.25% (RR) (RR) | (RR) (Rr) (Rr) (RR) | (rr) (rr)
6.25% (RR) (RR) | (rr) (Rr) (Rr) (rr) | (rr) (rr)

S₃

○ round = 62.5%

RR or **rr** homozygous = 75%

Gametes — R R | R r | r r

S₄ — (RR) | (RR) (RR) (Rr) (rr) (rr) | (rr)
16 | 8 4 8 4 8 | 16

S₄

○ round = 56.25%

RR or **rr** homozygous = 87.5%

(b) Open pollenation (random mating) after crossing of two homozygous parents

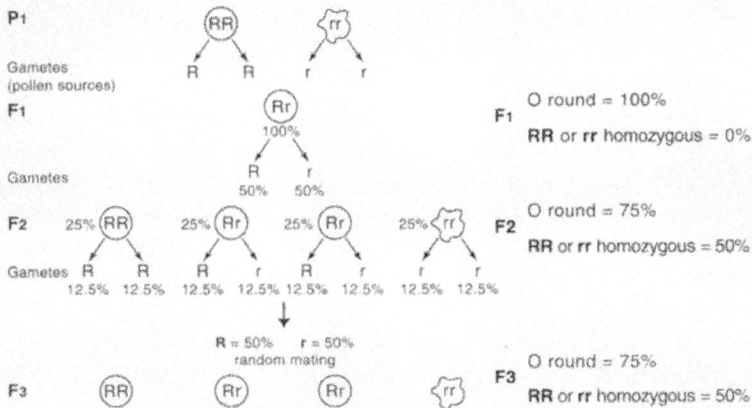

P₁ — (RR) (rr)

Gametes (pollen sources) — R R r r

F₁ — (Rr) 100%

F₁

○ round = 100%

RR or **rr** homozygous = 0%

Gametes — R 50% r 50%

F₂ — 25% (RR) 25% (Rr) 25% (Rr) 25% (rr)

F₂

○ round = 75%

RR or **rr** homozygous = 50%

Gametes — R R | R r | R r | r r
12.5% 12.5% | 12.5% 12.5% | 12.5% 12.5% | 12.5% 12.5%

R = 50% r = 50%
random mating

F₃ — (RR) (Rr) (Rr) (rr)

F₃

○ round = 75%

RR or **rr** homozygous = 50%

Figure 1.2 A plant splits off two female and two male gametes: crossing makes 2 × 2 combinations of gametes possible. After the crossing of a homozygote recessive plant (rr) and a homozygote dominant plant (RR) all offspring are the same genotype (Rr). (a) Self-pollination over generations will reduce the number of heterozygous individuals. After 6–9 generations 90–95% of the genotypes are homozygous for the trait. (b) Open pollination with at random mating and no selection pressure results in a constant percentage of the different genotypes.

A self-fertilizing plant that is homozygous for a particular trait will produce offspring that are also homozygous for the trait; the genotype is said to be 'fixed' or 'breeding true-to-type' (see Box 1.3). The group of identical homozygous genotypes is called a 'line'.

In cross-fertilizing populations the fractions of alleles remains constant in a stable situation, since in each generation pollen grains combine with any ovule (Figure 1.2b). (This constant relationship is called the law of Hardy-Weinberg.)

W. van der Heide, CAH Dronton

In the case of vegetatively propagated crops, the genotypes are 'fixed' since there is no fertilization, i.e. no recombination of genes through sexual propagation. Evolution is in this case strongly based on mutations. However, in most vegetatively propagated crops, such as sweet potato and potato, seeds may develop. Most of these vegetatively propagated crops are heterozygous for most characteristics and cross-fertilizing when flowering. Each seed from a cross thus represents a new, unique genotype. The plant growing from a seed can usually easily be vegetatively multiplied, producing genetically identical offspring.

The frequency of the alleles which make plants 'successful' in multiplication will increase in frequency in the population. Similarly, alleles which make plants unsuccessful will decrease in frequency. In the case of self-fertilizing and vegetatively propagated crops this means that the successful lines will make up a larger proportion of the total population. In cross-fertilizing species, the percentage of plants or genotypes carrying the successful allele will increase.

When the frequency of a particular allele in a population is low, i.e. few plants in the population carry the allele, there is an increased possibility that this allele may be lost by chance. This chance is significantly larger in a small plant population than in a large one. This random loss of alleles is called 'genetic drift'.

Farmer seed selection has a decisive effect on the frequency of genotypes in a landrace. If, for instance, a particular line is not well adapted to the environment because it lacks disease resistance, it will contribute fewer seeds to the total harvest. If the farmer takes a random sample for planting next season, fewer seeds of the susceptible genotypes will be sown. A farmer could also select pre-determined proportions of, for instance, red and white seeds from the harvest, regardless of their contribution to the harvest. If the red seed colour is associated with drought tolerance, the natural selection against drought tolerance in wet years would be offset by farmers' selection.

1.1.3 Centres of crop origin and diversity
In the nineteenth century travellers such as Von Humboldt, De Candolle and Darwin investigated the distribution of plants and animals over the

10

Box 1.3 Different mating systems, different seed systems

The mating system in plants refers to the manner the pollen, i.e. the male gametes, are deposited on the stigmas, i.e. the female parts of the flower and the method of fertilization. A flower forms pollen, i.e. male gametes, in anthers. The female gamete or egg cell is usually called an ovule. The ovule is usually surrounded by the tissue of the mother plant, with a style and stigma. In the case of open pollination, the pollen release from the anthers and the deposition of the pollen is not controlled. After pollination, the pollen germinates on the stigma and grows down the style to fertilize the ovule or egg. Successful pollination results in fertilization. The pollination can result in self- or cross-fertilization, depending on whether the pollen fell or was deposited (by man, an insect or the wind) on the stigma and fertilizing the ovule of the same plant or on the flower of another plant.

Self-fertilizing crops are crops in which under natural conditions 95 per cent of the pollination is 'selfing', that is, 95 per cent of the flowers are pollinated with pollen from the flower itself or with pollen from other flowers on the same plant. Modern varieties of self pollinators consist of one homozygous pure (geno-) type. Farmers can easily maintain the uniform genetic composition of the variety. Examples are: wheat, finger millet, bean, cowpea, groundnut, chickpea and tomato.

Cross-pollinating crops are crops in which under natural conditions cross-fertilization occurs in more than 50 per cent of the cases; pollen is brought from one plant to another by insects or by wind. Selfing in cross-pollinating crops often results in in-breeding depression, which expresses itself in a general weakening of the plants. Natural crossing may result in changes of the variety. Isolation (see Section 8.2) is a major concern for seed producers; variety maintenance is a continuous concern for breeders. Farmers cannot easily maintain varieties true to their original characteristics and may have to purchase seed regularly if they want to produce relatively uniform varieties. Cross-fertilizing crops, however, do have the capacity to respond to changing conditions, i.e. landraces of cross-fertilizing crops are more dynamic than those of self-fertilizers. Modern varieties are either composites, synthetics or hybrids (see Box 1.4). It is very important to know the type of variety when trying to maintain or improve them. Examples are: pearl millet, maize, sunflower, oilseed rape, cucumber and onion.

world. Based on the diversity of plant types found, the Russian geneticist-agronomist Vavilov (1887–1943) defined eight 'centres of origin' in the world for agricultural crops and their wild relatives (see Figure 1.3). Most important crops were indeed developed there from wild plants by farmers, and show important diversity. With present sophisticated methods of looking at genetic diversity, such as isozyme analysis, it has become clear that it is not true that most genetic diversity in a crop is necessarily found in its centre of origin. In areas outside the centres there

Box 1.4 Hybrids

Hybrids are a special case. A hybrid seed combines the characteristics of two parents. The parents are chosen so as to give the best combination, the greatest hybrid vigour. Hybrid seed has to be reconstituted each generation by crossing those parents, while avoiding self-pollination.

The seed produced on the hybrid, e.g. when a farmer tries to produce seed using a commercial hybrid, no longer carries the same features. The material loses most of the hybrid vigour and uniformity in the field. Hybrid seed cannot therefore be used in supporting local seed systems. However, when the parents are not very different, and where growing conditions are not optimal, this so-called F_2 seed has been reported to be planted by farmers quite successfully.

Farmers generally cannot produce hybrid seed themselves because the parent lines are normally not available. Hybrid varieties are therefore the main asset of the seed company producing the hybrid seed. Avoiding self-fertilization in hybrid seed production may be done in different ways: emasculation, male sterility or incompatibility.

In a number of countries, the word hybrid is used inappropriately in commercial promotion. Hybred, Hibred, Highbred or similar words are used to give a quality label to seed of an open-pollinated variety of crop, such as wheat, of which commercial hybrids are not produced.

Heterosis
A hybrid is the product of a cross between genetically unlike parents. In contrast to in-breeding depression, the cross-breeding of two different lines or cultivars will usually show more vigorous hybrid offspring than either of the parents considered separately. This superiority of the hybrid is known as heterosis. It shows itself in improved general fitness characteristics, such as longevity and resistance to biotic and abiotic stresses. This improved fitness is called hybrid vigour. Heterosis is quantitatively defined as the superiority of the hybrid over the average of the two parents.

Hybrids between in-bred lines in maize have dominated commercial breeding because of their marked positive effect on yield and guaranteed financial return to breeding companies. More recently also hybrid formation in field crops such as rice, wheat, sorghum and pearl millet have received increased attention. Hybrid breeding, to be successful, requires much attention and is more complex and expensive than population breeding.

We still know surprisingly little about the causal factors of heterosis at the biochemical-physiological level. The quantitative genetics have generally been well-established, although scientists still dispute each other's more detailed theories.

Figure 1.3 Centres of crop origin and diversity

1. The Chinese Centre: Soybean (*Glycine max*), Adzuki bean (*Phaseolus angularis*), Leaf mustard (*Brassica juncea*), Apricot and peach (*Prunus spp*), Orange (*Citrus sinensis*), China tea (*Camellia* (*Thea*) *sinensis*).

2. The Indian Centre: Rice (*Oryza sativa*), Finger millet (*Eleusine coracana*), Chickpea (*Cicer arietinum*), Egg plant (*Solanum melongena*), Taro yam (*Colocasia antiquorum*), Cucumber (*Cucumus satius*), Tree cotton (*Gossypium arboreum*), Jute (*Corchorus olitorius*), Pepper (*Piper nigrum*).

2a. The Indo-Malayan Centre: Yam (*Dioscorea spp*), Pomelo (*Citrus maxima*), Banana (*Musa spp*), Coconut (*Cocos nucifera*).

3. The Central Asiatic Centre: Bread wheat (*Triticum aestivum*), Rye (*Secale cerale*, secondary centre), Pea (*Pisum sativum*), Lentil (*Lens culinaris*), Chick pea (*Cicer arietinum*, one of the centres), Sesame (*Sesamum indicum*, one of the centres), Safflower (*Carthamus tinctorius*, one of the centres), Carrot (*Daucus carotus*, one of the centres), Apple (*Malus pumila*), Pear (*Pyrus communis*).

4. Near Eastern Centre: Durum wheat (*Triticum durum*), Bread wheat (*Triticum aestivum*), Einkorn (*Triticum monococcum*), Barley (*Hordeum vulgare*), Chick pea (*Cicer arietinum*, one of the centres), Rye (*Secale cerale*), Pea (*Pisum sativum*), Lentil (*Lens culinaris*), Blue alfalfa (*Medicago sativa*), Sesame (*Sesamum indicum*, one of the centres), Melon (*Cucumis melo*), Grape (*Vitis vinifera*).

5. The Mediterranean Centre: Durum wheat (*Triticum durum*), (hulled) Oats (*Avena strigosa*). Broad bean (*Vicia faba*), Cabbage (*Brassica oleracea*), Olive (*Olea europea*), Lettuce (*Latuca sativa*).

6. The Abyssinian Centre: Durum wheat (*Triticum durum*), Emmer (*Triticum dicoccum*), Barley (*Hordeum vulgare*), Cicer arietinum (*Chick pea*), Pea (*Pisum sativum*, one of the centres), Lentil (*Lens culinaris*, one of the centres), Flax (*Linum usitatissimum*, one of the centres), Sesame (*Sesamum indicum*, one of the centres), Teff (*Eragrostis tef*), Finger millet (*Eleusine coracana*), Coffee (*Coffea arabica*).

7. The South Mexican and Central American Centre: Corn (*Zea mays*), Common bean (*Phaseolus vulgaris*), Pepper (*Capsicum annuum*), Upland cotton (*Gossypium hirsutum*), Sisal hemp (*Agave sisalana*), Squash, Pumpkin, Gourd (*Cucurbita spp.*).

8. South American (Peruvian-Ecuadorean-Bolivian) Centre: Sweet potato (*Ipomoea batatas*), Potato (*Solanum tuberosum*), Tomato (*Lycopersicon esculentum*), Lima bean (*Phaseolus lunatus*), Sea island cotton (*Gossypium barbadense*), Papaya (*Carica papaya*), Tobacco (*Nicotiana tabacum*).

8a. The Chile Centre: Potato (*Solanum tuberosum*).

8b. Brasilian-Paraguayan Centre: Manioc (*Manihot esculenta*), Peanut (*Arachis hypogaea*), Rubber (*Hevea brasiliensis*), Pineapple (*Ananas comosus*).

13

is a wide genetic variation and in some places even more than in the area where the crop originated. These areas are called secondary centres of diversity. As more evidence of genetic diversity in crops became available, the definition of centres of origin and diversity have been further adapted and modified.

The local seed systems in the centres of diversity remain important sources of valuable genes for e.g. pest and disease resistance, because the crops and their wild relatives have been grown there for a long time in the presence of the pathogens. It is not, however, always the case. Potato varieties from the Andes do not have resistance to the fungus causing late blight which originates from the valley of Toluca in Mexico. In the Andes, the disease is relatively new for the crop.

1.1.4 Local seed systems

Repeated production and selection under local conditions, allowing for the effects of mutations, hybridization and selection pressure, represents a dynamic evolutionary process. In this system, on-farm seed production and local crop development are integrated with normal crop production. A combination of this on-farm system with seed and variety exchange among farmers forms what we call the local seed production system.

In short, local seed systems may be defined as systems in which selection, seed production and seed exchange are integrated into crop production and the socioeconomic processes of farming communities.

The term 'informal seed supply system' essentially has the same meaning as local seed system or farmers' seed system, emphasizing that the farmers' seed production activities take place outside the formal system of regulated seed production by registered organizations or institutions.

Surveys indicate that traditional farming systems foster high levels of genetic diversity. Farmers have many reasons to maintain and utilize diversity: (i) agronomic criteria (yield, yield stability, resistance to pests and diseases, storability, etc.), (ii) economic criteria (market and home economic) and (iii) cultural criteria. In selecting seeds and varieties, farmers combine and balance these criteria. As a result, farming systems in marginal, low-input areas tend to include more crops and cultivars than in areas where market integration is advanced and higher input levels are used. It should be borne in mind, however, that low-input small-scale production systems with low levels of genetic diversity also exist because of limited access to new materials and poverty (see Section 2.4).

1.1.5 Local varieties for multiple uses and preferences

The multiple needs of consumption and other purposes is one reason for farmers to use a range of crops and varieties. In Central America, different bean varieties offer the household some variation in the daily consumption of beans and maize. In the Andes, different potato varieties are used for fresh consumption (boiled), drying or freeze-drying, and for serving in different dishes. Some sorghum varieties in Africa are used for porridge (*sadza* or 'millies') and others for beer-making. Particular local varieties are grown for very special dishes, prepared for traditional cultural events. Special rice varieties are grown for special festival dishes in South-east Asia. A small white bean in central Costa Rica is made into an Easter porridge. Early maturing local varieties with names such as 'hunger conqueror' or 'poor man's harvest' may be planted to supply food early in the season. Very often local varieties are grown for home consumption, because they are better tasting. The modern, higher-yielding, but less tasty, varieties may then be grown in the same farms for the market. Particular varieties may be planted to provide building materials, brooms, fuel, dyes or medicine.

Many crops serve more than one purpose. Maize and barley produce grain for consumption and leaves and stems for animal feed. Some sweet potato, cassava or cowpea varieties are preferred because their leaves make a good vegetable.

Other reasons for making use of genetic diversity may be linked to beliefs or religion. Red maize is planted in maize fields in Mexico to safeguard the crop against bad harvest. In the Andean culture it is believed that the maintenance of a diversity of potato varieties supports prosperous yields. Farmers may maintain a range of varieties, because they differ in taste or because they may find some of the varieties useful for the future. This seems to be the case particularly (but not only) in vegetatively propagated crops such as in the Amazon basin of Peru, where an average of 12 cassava cultivars were found per garden. More than 20 varieties can be found in traditional potato fields in the Andes.

1.1.6 Local varieties for variable environments

Risk and yield stability
The diversity of crops and varieties also serves the stability of farm production. Stability of the yield is sometimes more important than the yield level. This concern is obvious when considering the variable production conditions faced by many small-scale farmers. The weather varies widely from one season to the next: one season may be wet, the next dry. A drought may be short or long, at the beginning of the growing season, in the middle or at the end. Disease pressure may be high in one season, insect pests dominating in the next. The market price is usually unpredictable, it can fluctuate between and within seasons. The farmer knows that every season is different, but does not know what to expect. His or her concern about food security for the family is obvious; the farmer cannot take the risk of producing below the minimum required for the family to survive.

15

Therefore, a variety which each season produces a reasonable, acceptable yield may be a better option for a resource-poor farmer than a variety with a high average yield, but with a greater risk of producing less than the required minimum.

Coping with variation in climate
One way to deal with a variable environment is to plant different crops and varieties in different fields ('between-field diversity') or mixed in the same field ('within-field diversity'). Such genetic diversity may 'buffer' against environmental variation. Because crops and varieties differ in planting date, date of flowering, length of the growing season, and suscep-tibility to pests and diseases, the risk that all are lost is reduced. This explains why farmers may want to plant maize and sorghum: maize is the best option for normal rainfall seasons, while in very dry years the sorghum can still give a reasonable yield. The decision to grow early and late bean varieties is similar: when the rains stop early, late varieties may wilt without producing any grain while the early ones are quick enough to mature. If the rainy season turns out normal, the late-maturing varieties will provide the highest yields. In areas where hailstorms occur, different planting and flowering dates reduce the risk of serious hailstorm damage of all crops or varieties.

Coping with variation in soil conditions
Growing conditions vary not only from season to season (variation in time) but also between fields (variation in space). Because fields vary in temperature and moisture regimes, farmers in Honduras plant hybrid maize varieties in the valley bottom and local varieties on the hillsides. One side of the valley may get more wind and sunshine than the other, which influences the choice of which crops and varieties to plant. Soils are another source of variation: sandy soils have different temperatures, nu-trient and water availability than clay soils. Within a field, growing condi-tions may be affected by variation in the inclination of the field, the presence of trees, boulders, or a river bench. Within-field variation is especially pronounced in marginal, infertile fields where plant growth is always limited by environmental constraints. Farmers are often aware of such variation, consciously or not. The way farmers use genetic variation to match environmental variation, rather than trying to make the en-vironment uniform, is illustrated by the case of the Mende farmers in Sierra Leone (see Box 1.5).

While variation in soils is one of the reasons for farmers to grow different crops and varieties, having a mixture of soils also helps to 'buffer' against the variation in weather conditions. In the very wet season, plants may suffer because of water-logging on the heavier soils and lower fields, while others give reasonable yields on the sandy soils with good drainage. In dry

Box 1.5 Farmers' adaptation of rice varieties to different environments, Sierra Leone

Mende upland rice farmers in Sierra Leone do not farm fields, they manage the complex cycle of vegetation changes associated with forest regeneration in systems of bush fallowing.

Since not all land is ready for planting at the same time, these techniques help minimize labour bottlenecks at planting and harvest, and maximize the period during which rice is available in the field. Short-duration rices ripen in 90 to 120 days planted on moisture-retentive soils at the foot of valley slopes, or in marshy depressions (*bului*) of the kind once favoured by elephants, serving to reduce the period of pre-harvest hunger. Medium-duration higher-yielding varieties with strong root systems are planted on free-draining upland soils when the rains are well set in (these continue today to supply the bulk of the crop). Long-duration types, ripening in five to six months and adapted to variable flood levels (including floating types) are planted in the beds of water courses and in inland valley swamps during breaks between the main upland operations, and are harvested at leisure after the main crop of dryland rice has ripened. Traditionally, Mende farmers do not irrigate: they match their planting strategies to the changing availability of soil moisture on the soil catena throughout the year. They follow the flood.

In effect, then, in labour-constrained circumstances, Mende rice farmers discovered that it makes more sense to develop the 'software' possibilities (manipulate germ plasm) than to rebuild the 'hardware' of agriculture (level fields and reshape the land for water control, etc.). How was this 'software' approach first worked out? Rice is largely a self-pollinating crop, subject only to small amounts of accidental out-crossing. The simplest method of harvesting, breaking off panicles one-by-one by hand, gives farmers the option to reject off-types as they harvest. Avoiding off-types in the course of panicle harvesting is equivalent to the procedure plant breeders term 'mass selection'. Panicle harvesting stabilizes the main seed types and also brings about a systematic grouping among off-types: early ripening types will be rogued as they ripen by farmers anxious to secure a little extra consumption in the hungry-season, and longer-duration types will be left in the field to the gleaners. In this way, over a long period of time, panicle selection has resulted in the differentiation of Mende rice germ plasm into three distinct duration classes.

Farmers are explicit about the need to maintain the pool of germ plasm biodiversity for rice through such experimentation and exchange. Informants have told me 'it is the nature of rice, and circumstances, to change'. No farmer believes in an ideal variety to which he or she will remain committed for life. The generations of rice are like the generations of humans: no child is an exact copy of their parent. Rice changes, however, within a framework governed by ancestral forces, the guardians of the moral order. It is through ancestral blessing that familiar and much-loved forms reassert themselves from time to time. In Mende this is given explicit recognition in the rice category name *mbeimbeihun* (literally 'rice-within-rice').

Source: Kate Longley, in de Boef *et al.*, 1993.

17

years the heavy soils and lower fields may have sufficient stored water reserves for a crop to mature normally, while crops on sandy soil wither.

Landraces and genotype mixtures
A special case of using genetic diversity is the case of genetically hetero-geneous landraces, in which *within*-variety genetic diversity may stabilize production. The reasoning is that a landrace that is well adapted to the location-specific variation, contains different genotypes: some genotypes do relatively well one season, and other genotypes do better when the conditions are different next year. Or some genotypes can survive on the salty parts of the field, while others do well on wet spots (see Box 1.6). Genotype mixtures may also vary in disease resistance which can limit the build-up of pest and disease. In special situations, mixtures may make more efficient use of water, light and nutrients. This has been shown for barley landraces in the Middle East: in some years early night frosts reduce par-ticular genotypes, and the remaining ones can utilize the water in the remaining part of the growing season.

It needs to be said, however, that information on the performance of landraces is fairly unsystematic. There is little evidence that stable yield levels of landraces are attributable to genetic variation within the variety.

1.2 The development of formal seed-supply systems

1.2.1 Industrialized countries

Increased understanding of genetics
In local seed-production systems as described in the former section, seed production and crop improvement are integrated, usually carried out on the same farm. Through the advances of science and technology, variety improvement and seed production have moved away from the farmers and have become specialized activities.

The role of farmers in crop development and seed production changed when the genetics of plant reproduction became better understood. In the middle of the nineteenth century, Mendel experimented with peas (see Box 1.2). His experiments remained unnoticed, however, until 1900. The rediscovery of his findings and subsequent studies developed into the science of genetics. Plant breeding became the practical application of that science. From then onwards, crossing and selection continually produced the desired varieties.

Technology and changes in agriculture
At about the same time, agricultural technology started to change rapidly in North America and Europe. In Germany, von Liebig found that plants

need minerals to grow and that the availability of more minerals makes plants grow stronger. Until that time the only way to fertilize crops was the application of animal dung or plant material (leaves, straw). The technology of large-scale, industrial fixation of nitrogen from the air produced chemical fertilizers. Mechanical technology developed fast in the early twentieth century: tractors and harvesting machines became commonplace farmers' tools. Crop and product uniformity became important to allow mechanized crop husbandry. Finally, chemicals were developed against a wide range of diseases and pests, allowing narrower rotations with fewer crops.

The combination of these new agricultural technologies, combined with increasing labour costs resulted in a rapidly changing agriculture after World War II in industrialized countries. Breeders at universities, government research institutes and private companies developed varieties that fitted these new technologies and conditions. By using insights in genetics and crop physiology in combination with a wide range of germ plasm, they were able to develop varieties which were higher yielding and responsive to fertilizer applications. Genetically uniform varieties met the requirements of crop uniformity. Disease resistance became less important because of the possibility of chemical crop protection.

Variety development and seed production as specialized activities
The new insights in genetics and breeding significantly advanced traditional farmers' crop improvement, which was principally based on selection. Thus, breeding, seed production and other agricultural technology became issues in national agricultural research and commerce. It is claimed that this change has reduced the access and control of farmers over genetic resources. Farmers now pay for inputs such as seeds that were developed from materials originally maintained by the farmers themselves (see also Section 4.2). The development of hybrids in maize and vegetables is sometimes considered an 'invention' by the private companies to secure their commercial interests, because farmers cannot reproduce this type of seed. Some claim that the potential of hybrids is based on maximum heterosis, while others think that had similar investments been made in the development of open-pollinated varieties, these could match the potential of hybrids. For commercial companies it is of course not logical to invest large amounts of resources in a product that farmers can multiply easily.

The development of a formal seed system: a chain of activities
With the changes in variety improvement and seed supply, a chain of specialized organizations developed: the formal seed supply system. In this chain, seed passes from one organization to the next, each organization being specialized and responsible for one or more aspect in the chain. At the beginning of this chain is the collection of germplasm in the farmers'

19

Box 1.6 The effect of a variable environment on a landrace: a hypothetical case

A variety or a variety mixture with different genotypes changes its genetic composition according to environmental variation. This is the basis for genetic adaptation. It can be explained by differences in yields between the genotypes in the variety.

The four presented figures are hypothetical, greatly simplified situations in which a seed lot of a variety made up of 3 genotypes (genotypes A, B and C) is planted and harvested over a number of seasons. Each figure shows the yield over a number of seasons and the presence of each of the genotypes in the seed lot. Each situation starts in year zero with the same seed lot: with 1/3 of the seeds of genotype A, 1/3 of genotype B and 1/3 of genotype C.

In situation 1, all following years are 'normal'. Because the genotype A yields most, it makes up the largest part of the yield, i.e. it contributes most to the harvest. As a consequence, if the farmer selects the seeds for planting next season at random from the harvest (that is without looking whether selecting genotype A, B, C), the proportion of seeds of genotypes A will automatically increase. The seed lot will also yield slightly more each season as the participation of lower yielding genotypes in the crop will slowly reduce over time.

In situation 2, disease pressure is high in all years. The yields are lower than in the normal situation. Genotype B, which is reasonably resistant, will each year increase its presence in the seedlot, at the expense of genotypes A and C. Since B is the best yielding genotype in the situations with high disease pressure, the increase of genotype B in the seedlot is associated with a slowly increasing yield level. The variety develops adaptation to the disease.

Situation 3 is a more realistic situation. Years differ and are never all normal and diseases are unlikely to be equally important each season. In situation 3, years with serious drought, high disease incidence and normal years alternate. Yields vary accordingly. The presence of the genotypes A, B and C also vary from season to season; none of the genotypes comes to dominate in the landrace.

In situation 4, different from situations 1, 2 and 3, the farmer selects for planting next year a seed lot with 1/3 of genotype A, 1/3 of genotype B and 1/3 of genotype C. None of the genotypes comes to dominate the landrace and a very dry year or high disease pressure do not affect the composition of the seed lot in the following season.

A = the genotype with a high yield potential, but sensitive to diseases and drought, B = genotype which is intermediate yielding, reasonably disease resistant, but sensitive to drought, C = low yielding, sensitive to diseases, but drought resistant.

For calculation of the yields and proportion of genotypes in the seedlots, assuming each year 300 seeds were planted, and seeds always weigh 1 gram, the following yields were used:

Genotype A: 10 grams per plant in normal years, 4 grams in years with high disease pressure, 1 gram in dry years

Genotype B: 8 grams per plant in normal years, 7 grams in years with high disease pressure, 1 gram in dry years

Genotype C: 5 grams per plant in normal years, 2 grams in years with high disease pressure, 3 gram in dry years

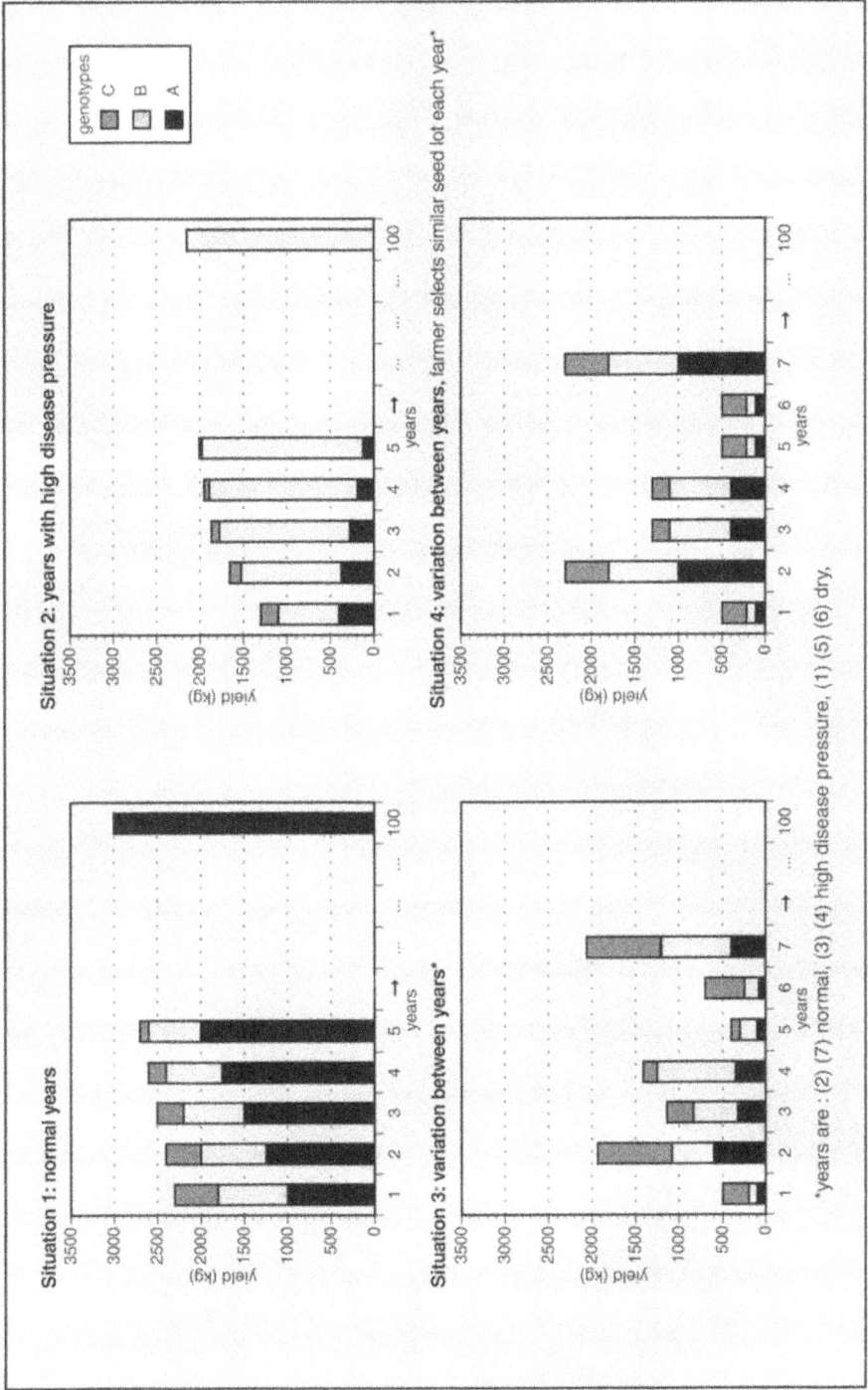

Figure 1.4

21

fields, thereafter maintained in genebanks. These collections consist of local varieties and wild relatives collected in traditional farming systems and 'old' modern varieties. The purpose of these collections was initially to provide breeders with useful genetic variation for their breeding activities. Lately the argument for long-term conservation has become a key issue (see Section 3.4). The next step in the chain is the breeding of new varieties and their maintenance by public and private programmes. Thereupon, successful new materials are multiplied by public or parastatal seed schemes, private seed companies or co-operatives. The following steps in the chain include seed distribution and finally the marketing of the seed to the farmers. Research, variety and seed legislation, and agricultural policy support this chain of activities. Such formal seed systems, in which all specialized activities depend on each other, are different from the local seed systems in which these seed-related activities are often carried out at the same farm, in an integrated manner.

The formal system provides the majority of seed in industrialized countries. However, figures indicate that, depending on the crop and the country, many farmers still produce and save seed for their own use. It is estimated, for example, that only 50 per cent of the total seed demand (over all agricultural crops) in Germany is seed produced by the formal seed sector. For Denmark, The Netherlands and Spain these data are 95, 75 and 10 per cent.

1.2.2 Developing countries

Variety improvement in developing countries
Changes in seed production and variety development in developing countries occurred later. Modern varieties became available from the early 1960s onwards through the breeding activities at the International Agricultural Research Centres. These varieties were high yielding in the high-input trial plots compared to the local varieties used by the farmers. They are therefore often called High-Yielding Varieties (HYV) or Modern Varieties (MV). In particular, the breeding programmes for rice (at IRRI), wheat and maize (at CIMMYT) have been very successful. For rice and wheat this success can be largely attributed to the development of short-straw varieties. Such short-straw plants divert more of their dry matter production into the development of grain. This increases the harvest index, the ratio between the weight of the harvested part (grain) and the total weight of the plant. In maize the development of hybrids has resulted in great yield improvements over the open-pollinated varieties.

Formal seed programmes and small-scale farmers
The implementation of a formal seed sector in developing countries according to the Western model has not been as successful as expected. Seed production and distribution programmes were implemented to distribute

the modern varieties to farmers. The programmes have in many cases been quite effective in more uniform favourable production environments, where modern varieties have had an important impact. The combination of these modern varieties with other inputs such as fertilizers and agro-chemicals triggered what became known as the Green Revolution. The Green Revolution has been most prominent in the large-scale adoption of modern rice, wheat and maize varieties in favourable production areas. This success has resulted in the disappearance of many local varieties, and in many situations in a reduction of genetic diversity (see Section 3.4).

Table 1.1: Percentage of area planted to modern varieties of rice, wheat, and maize in developing countries, 1970–90

	Rice[a]			Wheat[a]				Maize
	1970	1983	1991	1970	1977	1983	1990	1990
Sub-Saharan Africa	4	15	na	5	22	32	52	43
West Asia/North Africa	0	11	na	5	18	31	42	53
Asia (excl.China)	12	48	67	42	69	79	88	45
China	77	95	100	na	na	na	70	90
Latin America	4	28	58	11	24	68	82	46
All developing countries	**30**	**59**	**74**	**20[b]**	**41[b]**	**59[b]**	**70**	**57**

Adapted from Byerlee (1994)

a Excludes tall varieties released since 1965. If these varieties are included, the area under MVs increases, especially for rice in Latin America.
b Excludes China.

Although the programmes contributed significantly to the introduction of modern varieties, they proved in most cases to be not commercially viable. The structure and organization of these programmes were not suited to the reality of an important aspect of farming in developing countries: small and resource-poor farmers. The requirements in relation to the seed quality, timely delivery, distribution points, and so on, proved hard to meet. Furthermore, in the case of most important self-fertilizing food crops (wheat, rice, barley, beans) and vegetatively propagated crops (potato, cassava), farmers can easily multiply and genetically maintain a variety. Once they obtain seeds from a well-performing new variety, they buy new seed only in the case of crop failure or degeneration. The economic gains from using a higher seed quality usually do not justify the purchasing of seeds (see Chapter 9). Moreover, the seed is not always better quality than the seed reproduced by the farmer.

The situations in which hybrid maize and, to some extent, pearl millet have proved successful seed products may be considered exceptions. In such situations the commercial seed sector has often been more successful than the national breeding programmes. The same is true for (hybrid) vegetable seed.

Marginal, unfavourable environments
The formal seed sector has been particularly unsuccessful in meeting farmers' needs in less favourable and marginal production areas. Production conditions in these areas are usually more complex and more risk prone. Low soil fertility, frequently occurring droughts, inundations or other climatic hazards reduce the productivity of the crop. The availability or access to agricultural inputs (capital, land, labour, water, fertilizer, etc) are generally limited, more expensive and more variable (due to bad roads and a remote market). In these conditions farmers may lack cash to buy seed. Higher seed quality may not be expressed in higher yields because of other limiting factors, and the purchase of expensive seed may not be economical. Farmers in these areas may grow a wider range of varieties with different characteristics. Varieties that are suitable for favourable high-input conditions may not prove the best option for low-input conditions in these marginal areas. Moreover, modern varieties for the market may not meet the preferences for home consumption and the need for secondary products. Varieties that are bred for high yield may not serve the many different purposes for which a subsistence farmer produces a crop. The needs of farmers in marginal areas are usually quite varied and therefore difficult to address by formal seed programmes.

This does not, however, imply that modern varieties are never successful in less uniform and less favourable areas. In many situations modern varieties perform better than the local materials. They may have better pest and disease resistance and be higher yielding. They may be well suited for production for the market. Modern varieties may be added to the range of varieties already grown by farmers, partially replacing the older varieties and adding new genes to the local genepool.

Dependence on the formal seed system
Structural adjustment of the economy which has been carried out in many developing countries over the last decade has important consequences for the use of seeds and varieties. The privatization of the economy has resulted in the commercialization of seed programmes, the elimination of subsidies on farm inputs and food prices, and a lack of credit. A reduction of input levels seems the only option for many farmers. The replacement of expensive seeds of improved varieties by locally produced seeds implies, for cross-pollinated crops like maize, a return to local landraces. With lower seed costs and lower input levels, the planting of local varieties is often more economical than the use of purchased seeds of modern varieties. This situation has motivated many farmers and grassroots

organizations to engage in the re-introduction and seed production of local varieties, aiming at independence from the formal sector for plant genetic resources.

1.2.3 The use of local and modern varieties in local systems

The controversy about local versus modern varieties
Although a farmer may evaluate modern and local varieties similarly on their merits and their suitability to his or her farming system, there are some general features that can explain the arguments about local vs modern varieties. These general features are related to their origin: the conditions under which they were developed and the selection criteria of farmers and breeders. These general features have, however, resulted in a 'typical' characterization of the performance of local and modern varieties. There is, however, no clear systematic information that confirms the general difference in performance of these two categories of varieties.

Local varieties
The development of local varieties under natural and farmers' selection pressure into genetically heterogeneous plant populations was described earlier (Section 1.1). The character of local varieties is to a large extent determined by the farmers' selection criteria. This is usually clear for variety characteristics related to the use of the product, i.e. colour, taste, etc. The adaptation in terms of yield, yield stability and disease resistance to the environment and farmers' selection criteria is much more difficult to generalize. Because yield stability is an important aspect, natural and farmer selection is generally assumed to have been effective in creating populations with great yield stability, i.e. populations which produce an acceptable yield in all years (Box 1.7). Often, however, this yield stability seems related to low average yield levels. In situations where the local varieties have co-evolved with pathogens, there may be higher levels of resistance in the local materials, which can partly explain the observed yield stability. In other situations, resistance to pests and diseases in local materials is low. This is particularly true for crops and diseases outside their centre of origin (see Section 1.1.3). The local genepool may be limited compared to the pool of genes from which modern varieties were developed.

Modern varieties
The character of modern varieties is determined by the objectives of breeders. These objectives may result in modern varieties which may not fit the conditions of marginal, small-scale farming.

Public and private breeding programmes aim at a wide adaptation of their varieties: the more farmers can use a single new variety, the better. Breeders therefore test their materials in environments which together

25

Box 1.7 Yield stability

Yield stability relates to the variation in weight of the harvestable product (e.g., grains, tubers) under different growing conditions. If changes in growing conditions cause relatively small changes in production, yield is considered stable. If changes are relatively large, yield is considered unstable. This is similar to wide versus narrow adaptation, which is addressed in Box 1.8.

It is important to consider carefully the range of growing conditions in which the varieties are evaluated. The first important distinction to be made is variation in time versus variation in space. For instance, rainfall can differ greatly over years in sub-Saharan Africa, which leads to a wide variation in grain yield in a particular village. If grain yields in the same season differ among fields, villages, valleys or regions, there is variation in space. Of course, evaluations over, for example, two years in three villages will lead to variation in both time and space.

An aspect that is often concealed is the range in growing conditions that is under consideration. The examples of cultivars B1 and B2 in Figure 1.5a may represent landraces which are grown under low-fertility situations by farmers, and which are indeed stable under these conditions. However, the same stable landraces become unstable if evaluated at experimental stations; landrace B1 is not adapted to high-input conditions, whereas landrace B2 responds positively to the better growing conditions. The examples of cultivars C1 and C2 may represent official varieties that are bred at an experimental station, and that are not adapted to low-input conditions. However, they are stable under medium-input conditions. If growing conditions improve further, cultivar C1 responds negatively, and cultivar C2 responds positively.

A popular method of analysing yield stability is the Finlay-Wilkinson method (many more methods exist). A particular environment is not described by, for instance, the amount of rainfall, but by its average grain yield. Average grain yield of all germplasm is computed for each evaluation, and used as an independent variable. If, for example, germplasm has been evaluated at ten locations, there will be ten points on the X-axis that represent average grain yield at each of these locations (Figure 1.5b). Grain yield of one particular cultivar at all locations is used as a dependent variable (along the Y-axis) and related to average grain yield. The ten data points can be then described by a linear regression line, which is characterized by the point at which it crosses the Y-axis, and by the slope). It is the latter that quantifies (gives a value to) yield stability: the smaller the slope, the more horizontal the regression line is, and the smaller the change in yield due to changes in environment, and the greater the yield stability.

It is important, however, to combine this with yield level. Of two cultivars with similar yield stability, the one with the greater yield level would be preferable. Very often, a high yield stability is combined with a low yield level, while a low yield stability may be the result of low yields at low-input levels and high yields at high-input levels.

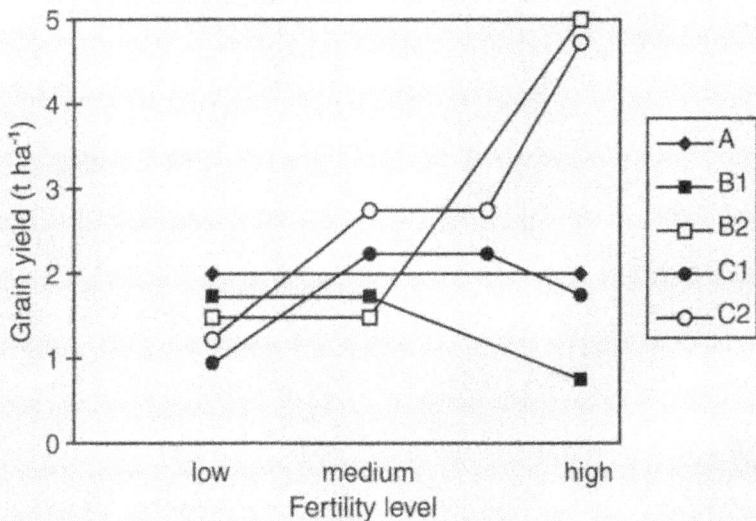

Figure 1.5a Yield stability

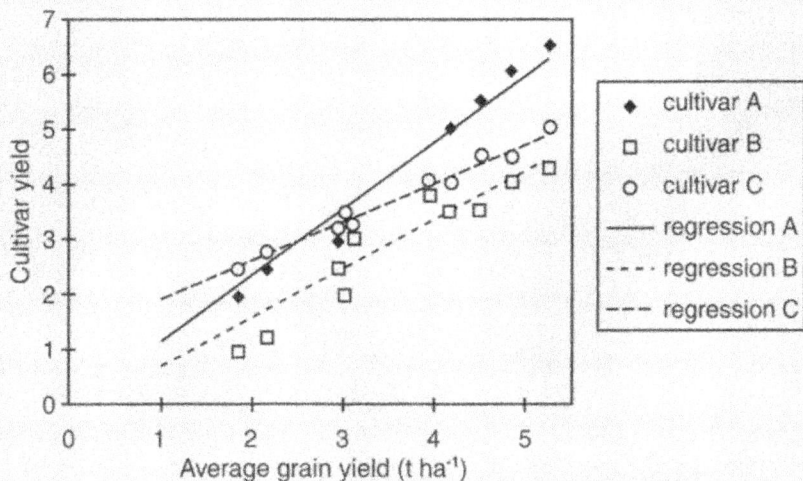

Figure 1.5b Yield stability

In a breeding programme it is carefully considered which environments are targeted. The wider the range of environments, the greater the yield stability of the material has to be. It has to be borne in mind, however, that variation in time may provide unexpected worse or better conditions, which germ plasm has to deal with in a positive manner.

Source: A. Elings, CPRO-DLO/CGN

represent the range of environments in their target area. From these multi-locational trials, breeders select the 'average best' variety.

The variety which performs well 'on average' may not be the very best choice in any particular environment. The conditions of the small-scale farmer may be quite different from the environment in which the breeder selected. Typically, breeders test their materials with larger fertilizer applications and with irrigation. In these conditions, the local checks may be out-yielded by the breeders' materials. When comparing the same materials under more marginal conditions, the farmers' local materials may out-yield the breeders' materials. This phenomenon is known as 'cross-over', an extreme type of 'genotype × environment interaction' (see Box 1.8). Genotype × environment interaction is common and cross-overs are found for a large number crops, but it is wrong to assume that these effects are always present or that they occur only when local and modern varieties are compared.

Another explanation for the unsuitability of modern varieties for small-scale farming is the fact that breeders may not be aware or are not able to address farmers' specific preferences. Taste, colour and secondary uses may be very important locally, but not relevant when considering the entire target area that the breeder is addressing.

Breeders develop modern varieties of self-pollinators that are genetically uniform, i.e. all plants are of the same genotype. In the case of cross-fertilizing crops like maize, breeders can produce either hybrids or OP's (open pollinated). In hybrids, all plants are close to identical, OP's are genetically heterogeneous. As compared to local varieties, it is probably fair to say that the genotypes in a modern variety are genetically more similar than in a local variety. From this it is often deduced that modern varieties are therefore less suitable for variable environmental conditions. There is, however, no proof for such an assumption in all conditions. Modern varieties can contain genes from materials originating from other parts of the world and may have had genetically very different parents. The combination of genes from very different populations in the world often produces heterosis (see Box 1.4). This can contribute to a good performance of modern varieties.

1.2.4 Linking the formal and the informal sectors

Two separate systems
The previous sections have described a common situation for developing countries in which two separate seed systems are poorly linked. There is a formal system, organized as a chain, in which the collection of germplasm, breeding, seed production and distribution are organized in logical sequential steps (Figure 1.7). This sector serves part of the agricultural production system, but is poorly equipped to meet the diverse needs of small-scale

Box 1.8 G × E interaction, narrow and wide adaptation

We speak of G × E interaction (genotype-by-environment interaction) if two or more genotypes respond differently to changes in the environment. To determine whether these responses really differ, we have to perform some type of statistical test.

Assume that without application of N fertilizer, cultivar A yields 2t ha^{-1} and cultivar B yields 3t ha^{-1}, and assume that with application of 100kg ha^{-1} of N fertilizer, cultivar A yields 4t ha^{-1} and cultivar B 5t ha^{-1}. Cultivars A and B therefore both respond to the increased availability of N with a grain yield increase of 2t ha^{-1}. As the increase in grain yield is the same for both cultivars, there is no interaction of the genotypes with the environments. The straight lines that describe the behaviour of cultivars A and B run parallel (Fig. 1.6a).

There is G × E interaction if there is a difference in increase in grain yields, for instance, an increase of 2t ha^{-1} for cultivar A and of 2.5t ha^{-1} for cultivar C. The lines that describe the behaviour of cultivars A and C do not run parallel and will cross at some point. This crossing over occurs beyond the range of grain yields observed (Fig. 1.6a).

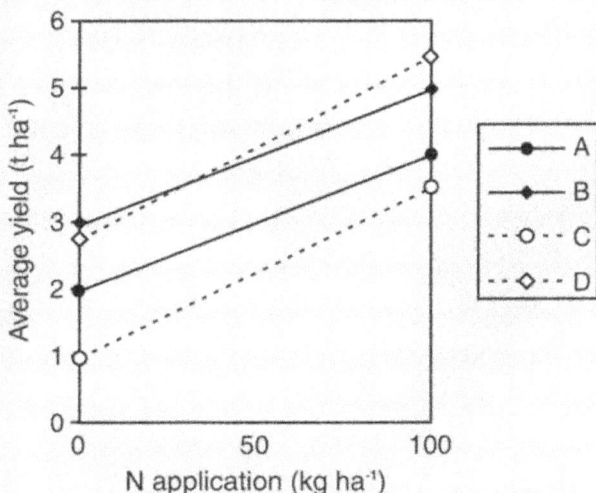

Figure 1.6a G × E interaction

We have to consider absolute values of the increase in yield; the fact that the relative grain yield increase of cultivar A was 100 per cent, and that of cultivar B 67 per cent, is not relevant. On the other hand, there is also G × E interaction with an increase of 2t ha^{-1} for cultivar A and of 2.75t ha^{-1} for cultivar D (from 2.75t ha^{-1} to 5.5t ha^{-1}), even though the relative increase in both cases is 100 per cent (Fig. 1.6a).

The issues of G × E interaction and adaptation are closely related. A cultivar has a narrow adaptation or is *narrowly adapted* if it performs well

(*continued over*)

under particular growing conditions, as is illustrated by cultivar 'narrow' in Figure 1.6b which yields well only with large amounts of N. If such a cultivar is grown at low N availability, it will fail. A cultivar that is *widely adapted* will perform reasonably well in many environments, as shown by cultivar 'wide' in Figure 1.6b which yields well under both low and high N conditions.

In Fig. 1.6b the yields of the two varieties show a crossover point around the 50kg N application. This means that at N-applications of less than 50kg the variety 'wide' yields better, but at higher N-applications, variety 'narrow' produces higher yields.

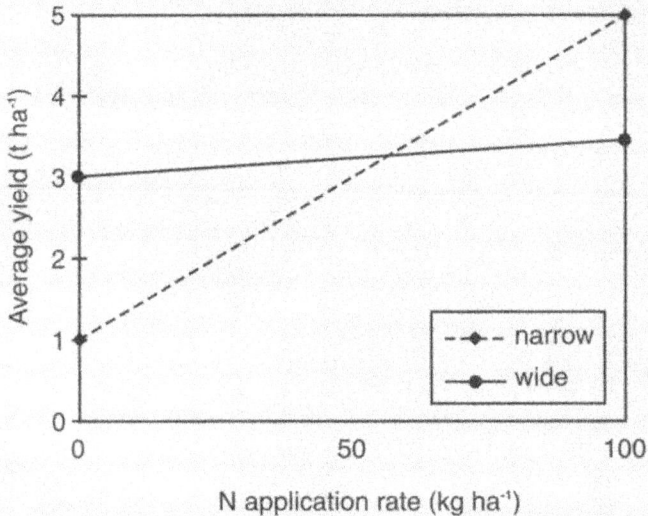

Figure 1.6b   G × E interaction

Source: A. Elings, CPRO-DLO/CGN

farmers. These farmers derive their seed from a local seed system. These local systems integrate production, genetic improvement and maintenance, seed production and exchange with crop production. Although there is some exchange of seeds and varieties between these two systems, they largely operate as two separate entities.

Complementarity of the local and formal seed systems
Interest in farmer seed production and the recognition of local seed systems is relatively recent. Studies of local seed practices have revealed that these local systems can be very flexible and dynamic. There are conditions where they are far better suited to small-scale farmers' needs in developing countries than formal systems could ever be. There is also a recognition, however, that in other aspects and situations, they have serious shortcom-

ings both in seed quality and (guaranteed) availability. This book is based on the assumption that both systems have much to gain by making use of each other's strengths. The integration of local and formal seed systems at the points where the systems meet can contribute significantly to the functioning of both.

Developing linkages
Developing such linkages requires an understanding of the local seed system. Such understanding enables strengths to be identified that can be used as a basis for support to these systems to overcome their weaknesses. Examples of such activities are the diffusion of seeds of modern varieties through traditional seed exchange mechanisms, and improving farmers' seed multiplication capacity. New approaches such as participatory breeding can contribute significantly to further integration of the two systems, building on each other's complementarity. This new approach requires recognition of farmers' capacity and the formulation of new goals for the formal systems. New goals for the formal system should allow the system to concentrate on the tasks it is best equipped for: breeding and maintaining a wide range of materials and the production of small volumes of good-quality seed, both for the modern sector and for feeding into the local

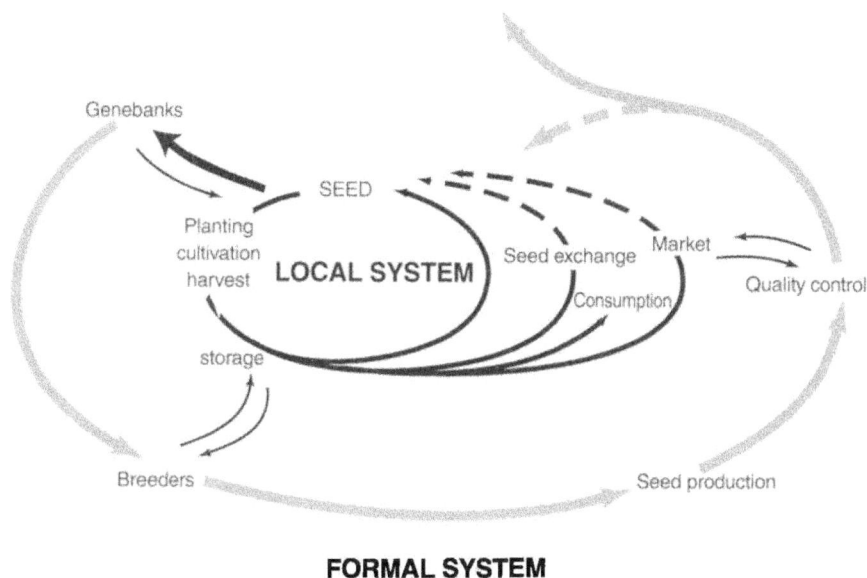

FORMAL SYSTEM

Figure 1.7 Two separate seed systems: the formal seed system as a chain of activities, and the local system in which different activities are integrated and locally organized, often on the same farm. The development of linkages between these systems (small arrows) will be of benefit to both systems.

31

system. It is also essential for those who work in the local sector to recognize the potential of the formal system and explore possibilities of support and technology from there.

The strategy to improve seed supply through strengthening the local system is not an isolated new approach. It corresponds to a new approach in other fields of agricultural technology development which take the local processes and the capacity of local people as a starting-point for development.

2. Improving local seed systems

The local seed system offers many opportunities for improving the seed security of small-scale farmers. Building on the strengths of the local system, options for the participatory improvement of varieties, seed production practices, the diffusion of varieties and the supply of seed are described.

2.1 The need to strengthen the local seed system

Local seed systems have proved to be dynamic and flexible in many aspects, but they also have weak points which require improvement.

The development of local varieties, their interaction with the environment through natural processes and farmers' selection provide examples of the effective processes of crop adaptation. Practices of local seed production and storage based on generations of experience are sometimes remarkable. Traditional seed exchange and farmers' knowledge are other valuable elements in many local systems. However, not all seed systems are characterized by these strengths. Furthermore, the world changes fast: increasing population pressure and the intensification of agricultural production require higher production levels, increased pest and disease resistance and adaptation to changing climate and soil conditions. We must recognize that evolutionary processes and farmers' knowledge alone cannot produce the changes which farming families presently need.

Building on the local system
Although there is pressure for change, this does not mean that traditional systems are of no use any more. The replacement of traditional systems by modern ones, developed for favourable conditions in the north, has not been a successful approach for many small-scale farmers in the south. Using the valuable characteristics of the local systems as building blocks for better, but still dynamic and flexible seed systems, complemented by improvements of the weaker points, is likely is to offer more sustainable options.

Because local production and seed systems differ from place to place, and the stronger and weaker points of the systems vary, this means that there is no blueprint according to which a local system can be improved. This is in contrast to the approach which has been using the blueprint of the successful 'Western seed system' for improving seed supply in developing countries (Section 1.2.2). Building on local seed systems requires analysis of the present situation, identification of strengths and weaknesses and a definition of strategies which can result in improved seed supply. This section will outline such strategies, based on the participation of the farmers in evaluating alternatives and options offered by 'scientific' or local knowledge, and by technology or seeds from the formal system or elsewhere.

Improving the local seed system means improving seed security, enhancing seed quality and the availability of good-quality seed. Improving seed quality involves:

○ the physiological quality (germination, vigour);
○ the health or sanitary status of the seed (absence of seed-borne diseases);
○ the analytical quality (percentage of good seed in a particular lot); and
○ genetic quality of the seed (varietal adaptation, varietal purity) (see Chapter 5).

These quality aspects can be improved through better seed selection, crop husbandry and storage practices. The availability of seed is improved by a general increase in seed quality and production levels. The activities are divided into three main categories:

○ variety improvement;
○ seed production; and
○ seed exchange and diffusion.

These categories broadly coincide with the three functions of a formal seed system: plant breeding, seed multiplication and distribution/marketing. They are further discussed in sections 2.2, 2.3 and 2.4.

Building on farmers' knowledge and capacities
In the past decade farmers' knowledge has been more widely recognized and valued as a resource for development. In many situations, farmers' practices and varieties have proved to be well adapted and the best option, given local climate, soils, the limited resources and diverse needs of the household.

Farmers have particular knowledge of their seeds and varieties. They are good selectors of varieties for their own use because they can weigh the different requirements at the same time: they can consider the needs of the household, how the variety fits into the total production system and how it adapts to the environment. Sometimes different groups of farmers within the community have different specialized knowledge or preferences (see Box 2.1).

Experimentation is a normal activity for farmers: most farmers experiment with new varieties if they have a chance. Their experimentation is not always obvious and does not work with conventional replicates and calculations. Farmers do not express their views in terms of plant genetics and cannot always describe why a particular variety is good. Their experimentation is, therefore, largely hidden. Group discussions serve often to bring

In villages near Cuzco, Peru, the native potato is considered a luxury: and they are highly valued even though they are relatively low yielding. The native potato is mostly grown by richer farmers in the community: the poorer cannot afford to grow them. It is the women who tend to know most about different native potato varieties, since they do most of the seed selection and look after the seed tubers during the storage season. Among the women it is the older and poorer farmers who know most about native potato varieties, and they hire their labour to the wealthier farmers in their community. They harvest and select native varieties in the richer farmers' fields (Zimmerer, 1989).

Women not only often have different expertise, they may also have different opinions than men on the value of contrasting varieties. For example, in a bean programme in Colombia, the bean varieties selected by the participating men were not the ones that were most favoured by their wives. The men selected the highest-yielding varieties, while the women preferred a variety with a small bean-type. The women, being responsible for the food and cooking, explained that they valued the small beans because more portions could be served with them.

(Ashby, Quiros and Rivers, 1989)

out such experiences with experimentation and are a good starting-point, therefore, for analysing the present situation and identifying problems.

2.2 Crop improvement: participatory plant breeding and variety selection

2.2.1 Introduction
Genetic crop improvement is an attractive way of increasing productivity. When seeds can be saved and used the following season, the benefit of better genetic quality of seed can be reproduced: varieties with a higher yield potential, with more efficient fertilizer use and better pest and disease resistance can produce better harvests over a number of seasons. This is the reason farmers are usually interested in variety improvement. The following sections address different forms of crop improvement in which farmers participate. Two forms of variety improvement are distinguished: (i) introduction of new materials; (ii) improvement of already available and used materials.

2.2.2 The introduction of new materials
In many situations farmers have only limited access to new genetic material: the formal or informal seed channel does not import exotic seeds, there is no information on new varieties, other seeds are not available or it is

unknown where they can be purchased. For the national researcher or technician, new varieties of seeds are usually available from the national breeding programmes, but means for the distribution and adoption of the materials among the farmers are limited.

Adoption of new varieties in the local genepool
A new variety may be a local variety from elsewhere, a modern variety or any advanced genetic material; in any case, it is material exotic to the local system. After evaluation and acceptance, a new variety may replace one or two other varieties, or it may be added to the existing range already exist-ing. If added to the range of varieties already grown, the new variety broadens that range and can make the production systems more diverse.

Adopted bean varieties in Central Africa or potato varieties are often added to mixtures of varieties, sometimes after evaluating them in monocul-ture. In other situations, a new variety may be maintained separately or a number of new materials combined to constitute a new mixture. The intro-duction does not necessarily affect the total number of varieties grown.

In the case of cross-pollination, genes from the new variety may be incorporated in the local materials (see Section 1.1). This can change the appearance of the local variety. Where such a gene improves the yield of the locally used variety, it is considered a genetic enrichment. When it negatively affects the local variety it is often called genetic contamination.

2.2.3 Improving and maintaining varieties
The repeated local production and saving of seed by farmers can change the genetic composition of the variety (see Sections 1.1 and 8.2). When this change is negative, it is called genetic degeneration. This can happen in both local and modern varieties. There are two ways of handling this: improving degenerated varieties or preventing the varieties from degener-ating, i.e. appropriate maintenance of the variety. In Chapter 8 this is referred to as improving or maintaining the genetic quality of seed.

Until recently, the improvement of varieties already used by the farmers did not receive much attention. Generally the aim of a seed project was to introduce improved varieties and the purchase of new seed by farmers for each planting season. However, experience indicates that local varieties also can be well-adapted and in the absence of better alternatives, genetic improvement can provide an interesting option.

There is plenty of evidence of farmers' capacities as breeder-selectors when looking at the level of adaptation of their varieties to local conditions.

An interesting example of farmers' seed selection was found in Tigray, Northern Ethiopia (Box 2.2). Farmers have selected and developed crops and cropping systems which are extremely well adapted to the particular growing conditions in their area.

Box 2.2 Mass selection: how simple?

When the professional literature of plant breeding mentions traditional practices, these practices are usually referred to as 'simple mass selection' and
sometimes further qualified with words such as 'crude' or 'primitive'. Standards of science demand more precision, but farmers' practices need to be
more accurately described in order to be fully understood. The following
examples from Ethiopia are illustrative.

Once an Ethiopian farmer explained to me how he selects chickpeas. He
took me to the field. 'Look here', he said. 'This plant is vigorous *relative to the
surrounding plants*. That will be selected.' This is a form of grid-selection,
more efficient than simple mass selection.

Another example of such an approach was given by a farmer who had
striga-resistant sorghum. He told me that he would always look for the most
severely striga-infested patch of land and look for candidates for selection
there.

In the highlands of Tigray, I found a two-year cycle of selection being
systematically applied to barley and wheat. Selected seed heads were
picked from the main field and planted in a separate field for propagation
next year. That field would be located on the deepest and most fertile land on
the farm. It would also be manured and given particular attention. The harvest would be bulked and planted in an ordinary field the second year. Then
the cycle would be repeated. This is a form of shuttle-breeding, shifting
populations between two different fertility environments. This selection system is called *mangas* (*ma'gas!*), meaning 'making it a king', or to 'ennoble', a
word occasionally used for plant breeding. Thus language reveals that traditional seed selection systems and modern plant breeding are perceived in
basically the same way. It is high time that these farmers were taken seriously and recognized for what they are: plant breeders.

Source: Trygve Berg, Plant Breeder/Scientist, NORAGRIC, Agricultural University of Norway, P.O.Box 5002 Aas, Norway In: *African Diversity* 13
(1994).

Variety maintenance and variety improvement have different objectives
but are closely related and involve very similar activities. In certain situations, maintenance and improvement are based on the selection of seed
from plants with particular defined, desired characteristics, eliminating the
less desirable ones (positive and negative selection). Variety improvement
and maintenance can be an objective in local and in modern varieties
(variety maintenance and improvement practices are discussed in Chapter
8).

A number of techniques are used in local seed systems to maintain
varieties. African rice farmers harvest seed from the centre of the field to
maintain the 'purity' of the rice and to avoid introgression from other

landraces and wild *Oryza* species surrounding the field. Sorghum farmers select heads for seed from a range of plant types. Mexican farmers manipulate hybridization between maize cultivars by varying planting dates and, consequently, the period of flowering of the different varieties. Tanzanian farmers separate hybrid maize fields from those of local maize in order to protect the local types. Farmers' seed selection in combination with production under local conditions has created local cultivars with valuable adaptation to specific local conditions, such as the day-length adaptation of sorghum cultivars in Nigeria and the drought resistance of maize cultivars in the south-west of Northern America.

As well as maintaining varieties, farmer selection aims at improvement. Examples of such local crop improvement can be found, for example, with maize farmers in Central America. They are aware of the fact that local and modern maize varieties 'marry' when planted close to each other. They may even favour such hybridization in order to obtain seed that combine the desired characteristics of both: ears which are well enclosed by husk leaves (characteristic of the local variety) and short plants which yield well and are less vulnerable to falling over in strong winds (characteristic of the modern variety). It is also reported that farmers in this region promote the crossing of peppers and weedy relatives in order to increase their spiciness.

Variety improvement and maintenance in cross-pollinating crops such as maize are more complicated than in the case of a self-pollinating or vegetatively propagated crop, and carry the risk of in-breeding when too few plants are selected (Section 8.2). This may have happened to many local varieties of maize, sorghum and other cross-pollinating crops.

2.2.4 Participatory crop improvement

Definitions
Plant breeding consists essentially of the generation of genetic diversity and selection from that diversity (Section 8.3). Participatory plant breeding aims to involve the farmers in the breeding process. In conventional plant breeding, farmers are usually only involved in the on-farm testing, i.e. the last part of the breeding process, when the range of genetic material the breeder initially started with has been reduced to 2–5 nearly finished, potentially successful varieties.

Two types of participatory breeding are now generally distinguished: Participatory Variety Selection (PVS) and Participatory Plant Breeding (PPB). Some consider PVS as a form of PPB, others see PVS and PPB as two different forms that exist next to each other.

Some people also differentiate between other forms of participatory crop improvement: 'breeder-led' for those situations in which the breeder leads the process and takes the decisions, and 'farmer-led' for those situations in which the farmer is the leading and deciding actor. Others distinguish between consultative and collaborative participation: the former indicates a situation in which the farmer is asked for information and advice but does not participate in the decision, as in the case of collaborative participation.

Participatory Variety Selection (PVS) is the term for the activities in which farmers evaluate and select from among released or pre-released or advanced (i.e. nearly finished) varieties. Farmers may be invited to the experimental station or to district variety trials to select and take seed from the varieties they consider promising. In another form, a group of farmers may implement a community evaluation trial, or farmers grow and evaluate a number of varieties on their own farm.

Participatory Plant Breeding (PPB) usually refers to the activities in which farmers select plants or seeds from and within a genetically variable population or variety. In the case of self-fertilizing crops such as beans or rice, farmers may work with unfinished varieties. These unfinished varieties, i.e. early generations segregate genetically when multiplied until finally lines are developed (see Section 1.1 and Box 1.2).

In the case of PPB with cross-fertilizing crops (maize, sorghum), farmers also work with genetically variable populations; a hybrid variety is the only case in which a cross-pollinating crop variety is made up of identical genotypes. The multiplication of cross-pollinated crops always recombines the genes and thereby changes the genotype make-up of the variety as a result of environmental and farmer selection (see Section 1.1 and Box 1.2). Using this definition, one could say that PVS for cross-pollinating crops does not exist: it also implies that a farmer who produces, selects and saves seed from a cross-pollinating crop practises a form of PPB. It does not however, imply that the farmer consciously does so or that his or her practices are effective.

PPB-like activities have so far principally addressed the PVS with self-fertilizing landraces; experience in the development of improved populations for further selection and maintenance by farmers is at present very limited. Of course, the maintenance of open-pollinated varieties (OPVs) of maize and sorghum provide such an opportunity. It is known that many farmers maintain OPVs of these crops over many years, but there has been surprisingly little study of farmers' selection practices and changes in such varieties.

In PPB, farmers may also be involved in the part of the breeding process where the genetic diversity is generated; the part of the process preceeding the selection. Farmers may be involved in the selection of parents for the crosses. Farmers can identify the local varieties of which they would like to have the characteristics combined with improved pest and disease resistance. Farmers can also participate in the actual crossing itself. Farmers in Mindanao, the Philippines, received training from an NGO in crossing rice, enabling them to make their own decisions in selecting and crossing parents and thereafter select from the generated diversity (Bertuso et al. in prep.). Such capacity to make crossings increases the potential access of farmers to genetic diversity.

The type of crops and the traits to be selected for are very important in the planning and design of PVS or PPB activities (see Chapter 8). Because selection within genetically variable populations is more complex and requires more of the farmer's time, usually over a number of seasons, the general idea is first to evaluate existing, available varieties or advanced materials through PVS-type activities before engaging in true PPB activities. PVS is relatively easy and can provide important short-term successes.

Reasons for farmers' participation in breeding
The principal reason for increasing the involvement of farmers in breeding is a better targetting of the environment and meeting the preferences of the farmers. Marginal, heterogeneous and remote environments are difficult to address through centralized formal breeding programmes. Varieties for these environments need adaptation to a specific combination of environmental stresses, to socio-economic conditions and specific preferences of taste and other characteristics. Because of the limited financial means, international and national breeding programmes have to set priorities in their objectives. To develop products which will be used by many farmers, they usually aim at developing varieties for high-input agriculture in more uniform, high-input areas (see Section 1.2.2). The results from these breeding programmes are a relatively small number of varieties that have high-yield potential in large, relatively uniform high-input environments. These varieties are often not adapted to the growing conditions or the needs of small-scale farmers in marginal areas.

These limitations of breeding programmes have stimulated the idea of transferring the final phase of a breeding programme to the farmers themselves (see Box 2.3). Instead of the breeder selecting over a number of seasons, mostly on-station, eliminating most materials and maintaining only the best performing ones for the farmer to choose from, farmers could select from a much larger number of options. These materials may sometimes include unfinished products which are not (yet) genetically stable.

Such an approach could:

○ Reduce the possibility that the breeder eliminates materials in earlier cycles of selection that are potentially suitable for particular conditions or meet the farmers' preferences. This would also mean that farmers have easier access to more genetic diversity.
○ Offer opportunities for developing better local adaptation: final selection is done *in situ*, by the users themselves.

Box 2.3 Participatory approaches in bean variety selection in Rwanda

Women farmers in Rwanda plant different bean variety mixtures for different conditions on their farms: different soils, sunny areas and areas under the shade of banana trees. They compose these mixtures through experimentation. A woman farmer may test 75–100 bean varieties in her lifetime. In Rwanda some 550 varieties exist countrywide. Despite this situation, the bean breeding programme, following Western models, screened 200 bean entries on station of which finally only 2–5 were selected. In exploring the options for improving the effectiveness of the bean programme, the question was raised whether and how farmers could absorb and use a much wider range of varieties than those provided by the formal system.

In a first phase of the project, women farmers, acknowledged in their community for their expertise in variety selection, were invited to evaluate 15 cultivars in on-station trials, 2–4 seasons before the on-farm testing. This evaluation resulted in the series of observations. Farmers used more selection criteria than breeders, and they combine some of these criteria. An example of such combined traits is the ability of a variety to perform well under a banana stand. This trait combines characteristics such as uprightness and sturdiness of stem, height of the lowest pod, and yield. It was also clear from the evaluation that the farmers could extrapolate from the on-station trial how a particular variety would perform in the conditions of their own farm, with a different soil fertility and crop-association. The selected bean varieties outperformed their own checks on their own farm with yield increases of up to 38 per cent, while the breeder choices in the same region showed insignificant gains. From the evaluation it was also clear that different farmers or farming communities had different needs and preferences. Differences in soil fertility on their farms can be factors explaining these differences. In total, 21 varieties were selected by the farmers in this phase of the project, which matches the number of varieties the breeding programme had released over the last 25 years.

In a second phase of the project, the potential for an earlier involvement of farmers was explored, with the possible benefits of reducing on-station trial costs, and speeding up the delivery of well-adapted varieties to the farmers. For three years, farmers were invited to view on-station trials containing up to 80 bean lines which were screened for the most important diseases in the region. The trials were planted without fertilizer and special management to enhance yields. In one season each variety was planted in a $3 \times 3m^2$ plot instead of a double row, to facilitate evaluation by the farmers. Initially, farmers were asked to evaluate each variety by assessing positive and negative traits, in an interview. Later, in moving the activities to the communities and aiming at decentralization, farmer-selectors representing their community were asked to mark the varieties they wanted to test on their farms with coloured ribbons. In the two first seasons of the second phase of the project, 26 varieties were selected for home-testing.

At this point, several institutional concerns were raised. It was not always the farmer with most expertise who was invited to represent the community: representation can be distorted by hierarchical power structures. As a result, community interests are not well served and, for instance, seed may not be distributed after multiplication. Also, the variables that indicate impact may need adaptation. Traditionally, the impact of breeding programmes is usually measured by the number of varieties released, and the rates of variety adoption. In participatory breeding trials, aspects of access to germplasm and enhancement of genetic diversity may also need to be assessed.

Adapted from: Sperling and Scheidegger, 1995 and Sperling and Ashby, 1997.

○ Contribute to increasing the genetic diversity in agriculture: farmers may select different materials, and genetically diverse populations may evolve differently in different environments.
○ Accelerate the introduction of promising materials into farmers' fields. Instead of carrying out the 7 or 10 seasons of selection on-station, the breeder may give the farmers materials from much earlier selection cycles.

The idea of involving farmers in variety selection has been strongly supported by the cases in which farmers managed to identify successful materials from on-farm experiments or demonstration plots. A good example is the diffusion of the rice variety Mahsuri. After two years of testing in India, the variety, which originated in Malaysia, was rejected by rice breeders in 1968 because of its lodging behaviour. Through a farm labourer, however, some seed found its way to the villages of Andhra Pradesh. The farmers found the variety excellent. As a result of the farmer-to-farmer seed exchange, Mahsuri became the third most popular rice variety, after IR8 and Jaya dwarf rice (Maurya, 1989). Because of the demand for quality seed and pressure from farmers, the Indian government was forced officially to release the variety, so that certified seed could meet the demand. Similar instances of the diffusion of seed via farmer-to-farmer exchange are reported for other crops elsewhere.

2.2.5 Examples of PVS and PPB
Examples of farmers' breeding activities and collaboration between breeders and farmers in selection are still limited. The cases of farmers practising forms of PPB are particularly difficult to identify and document. Most of these cases are anecdotal in the way they are described. There is, for instance, the case of farmers in the Philippines and Sierra Leone, who

are reported to have selected off-type plants from their field and multiplied the seed separately, generating their own varieties.

One of the first examples of PVS was reported in the work of scientists and farmers in Colombia (see Box 2.1). Other interesting examples of farmer-breeder collaboration are from Rwanda (see Box 2.3), Bolivia (Thiele et al., 1997), and Nepal (Whitcombe et al., 1996). Experiences with PPB are also reported from Brazil and India (Eyzaguirre and Iwanaga, 1996).

In Nepal PPB has been successful in producing cold-tolerant rice cultivars by involving farmers in the selection from an F5-bulk onwards (Sthapit et al., 1996). Cold-tolerance was not an important objective for the formal rice breeding programme in Nepal. The objective of the PPB project was to breed acceptable varieties with a minimum of resources by using farmers' knowledge. A local rice variety was used as a parent in crosses with improved materials. The local variety originated from India and had been used by farmers in Nepal for 25 years because of its cold tolerance, but they wanted to improve on its grain colour. From the crosses, the breeders selected lines in the F_4 generation on-station, multiplied these and provided farmers with samples of 20–25g of seed. Farmers were told about the segregation of traits in the population, as at the F_4–F_5 generations the lines are not yet genetically homogeneous (see Box 1.2). They were told that there would be no further segregation after selection over 2–3 years. Farmers selected for different characteristics: grain colour, yield, plant height and maturity relative to the local variety were the most important ones. Some farmers also considered panicle form, tillering ability, cooking quality and grain set. Most farmers selected a single plant from the well-performing crops and multiplied the seeds separately. In this way, farmers developed well-adapted cultivars which differed from farmer to farmer. Diffusion of these different materials increased biodiversity in the villages.

In Colombia, CIAT breeders compared their selection of F_2 populations of common bean with those of farmers, using the same, simple selection strategies to develop homogeneous lines (Kornegay et al., 1996). The farmers were told about the segregation of traits, but were invited to use their own criteria in making the selections. The time of selection and the number of plants per line selected was left to the farmers, but the breeders suggested to the farmer the number of lines to select to maintain a reasonable number of entries and force the farmers to use selection pressure. The breeders selected on station and the farmers on their own farm. After five seasons of selection, the best lines of each of the breeders and farmers were grown on the two stations and three farms. The results showed that neither the breeders' nor the farmers' selections were the best yielding in all fields. The breeders tended to have selected better yielders, but the farmers developed varieties with better bean colour.

Landrace enhancement is also a form of PPB. It aims to improve a local variety, while maintaining its typical, favourable characteristics. Landrace enhancement can aim at yield improvement, and increased disease resistance, while maintaining the genetic diversity which provides the local variety with yield stability. Landrace enhancement can be based on selection only (see Section 8.2). This can be the selection of seed from the best plants from a local variety (positive mass selection) and multiplication of the selected seed in a separate field. Another means of improvement is the elimination of the less desirable plants from a local variety (negative mass selection).

There are examples of relatively simple selection of the ears of the best plants of self-pollinating crops which resulted in considerable yield improvements in local varieties of rice (work of the Cantho University with farmers in Vietnam, part of the CBDC programme) and of barley and wheat (work of the Ethiopian Biodiversity Intitute in Ethiopia). There are also positive examples of selecting for shorter and earlier maturing plants in genetically heterogeneous local varieties.

For cross-pollinating crops, positive mass selection has to be practised with care. A pilot project at the Zamorano Panamerican School of Agriculture, Honduras, showed that selection of 25–40 plants in the farmer's field, and pollination between these plants by bulking the pollen from these plants and fertilizing their bagged ears, produced considerable yield increases in some local varieties after one and two years. However, repeating this procedure over several seasons, yields tended to decrease, which could be a result of inbreeding depression. The case illustrates well the difficulty of positive mass selection in cross-pollinating crops in farmers' relatively small fields and with the few ears a farmer normally selects. Two hundred ears of maize is considered the minimum to avoid such inbreeding effects. (see also Section 8.2.5).

In Brazil, Rio Grande del Sul, farmers with the support of an NGO (CETAP: Centro de Tecnologias Alternativas Populares) have developed a way to obtain maize seed which combines genetic improvement with seed production. The farmers cannot afford to buy the expensive hybrid maize seed every planting season. They found out that planting the maize hybrid together with their local variety produced a seed which yielded better than their local variety and was much cheaper than the hybrid (Gusson, 1998). Another successful product of PPB in maize is found in a community in the State of Rio de Janeiro. The farmers in this area grow maize on very poor and acid soils, under very high temperatures. A breeder from EMBRAPA, the national agricultural research organization, provided the farmers with a plant population developed by crossing 36 different materials, both improved and local ones (Toledo Machado, in prep.). Thus, the population contained ogenes from varieties which had tolerances to soil acidity, high temperatures and high yields. Farmers developed a new variety from this population by selection and called it 'Sol da Manhã (Sun of the Morning).

Because they are cross-pollinators, crops like maize and sorghum provide opportunities to incorporate pest or disease resistance into local varieties by planting them together with varieties which do have the required resistance, allowing free, open pollination and selecting resistant plants in the following seasons. No such examples have been documented yet.

The appropriate use of participatory plant breeding
Participatory approaches in plant breeding are currently receiving much attention. Such approaches are essentially not new and have been successfully used in developed countries such as The Netherlands (see Box 2.4), as well as developing countries (see Box 2.4). However, the potential of PPB for different crops, environments and farmers is still very much under debate. Not only the types of materials and selection practices, but also the institutions involved need to be of the type to ensure that more farmers than the few who participate in the pilot project will benefit. Furthermore, the benefit or impact of a PPB-approach must be clear for both the farmers and the institution.

Box 2.4 Participatory plant breeding: Dutch potato breeding

In recent literature, PPB is described as a completely new concept which will work particularly well with small and remote farmers in marginal conditions. In fact, even modern plant breeding in the Netherlands is very participatory.

Breeders of potato-breeding companies in The Netherlands, accept that particular farmers can select more effectively than they can. The company makes the crosses between particular parents. The seeds or the young clones are given to farmers who perform the main selection work over a number of seasons. The farmer subsequently sends the selected tubers to the company for specialized disease resistance screening, after which the farmer and breeder discuss the commercial prospects of the new selection. Promising varieties are registered and tested for entering the variety list and commercial revenues from the sale of the variety are shared between the farmer and the breeding company. Some farmers have earned a lot of money through this co-operation. This is a true form of participatory plant breeding, which is not particularly aimed at specific adaptation or low external input agriculture.

While the opportunities seem very promising, there are likely to be some bottlenecks in PPB. The following questions need to be taken seriously.

o If programmes for crop improvement for marginal environments are difficult and slow for breeders, will PPB be an effective approach for the farmers? PPB, like conventional plant breeding, is a long process, requiring long-term commitment from the participants. For the farmer to engage in a breeding-selection process over a number of seasons means a large investment in land and labour. The farmer's advanced variety may easily be lost because of droughts, flooding, diseases or robbery. The benefit for the farmer must justify these investments and risks. Therefore, the first option should be PVS, to look for promising varieties which are already available.

○ Which characteristics can be improved through PPB? Not all characteristics are equally suitable for improvement through participatory breeding. While some complex traits can be selected effectively for local conditions by one or two farmers, others may require statistically planned trials with repetitions. For other characteristics the more complicated selection schemes used in breeding programmes may result in the greatest improvements quickly. However, these schemes may not be very farmer friendly or easy to use.

○ How can farmers than other those who participate in the PVS and PPB benefit from the results? Or how can PPB be scaled-up to reach more farmers? When only key farmers engage in the selection part of the process, well-functioning seed multiplication and diffusion channels should exist for other farmers to obtain the improved varieties.

2.3 Improving seed production

2.3.1 On-farm production and storage of seed

Although local knowledge and farmers' selection ability may be impressive, there are usually opportunities to improve on-farm seed production. This section presents a general outline of the justifications for and the methods of working with farmers on improving their seed production potential.

Support to local seed production has to be preceded by an analysis of constraints. This analysis should yield the most pressing problems and not concentrate on those perceived by a seed technologist, based on the quality standards followed by the formal sector. Not all the aspects and indicators used in the formal sector are equally relevant for small-scale or low-input agriculture, e.g. the presence of stones or straw may not be a problem in farm-saved seed of a hand-sown crop, whereas it is a major quality aspect in packed commercial seed.

There are two fundamentally different ways to improve on local seed production:

○ improving the farm-saved seed (of all farmers in the target area); and
○ improving the seed production of selected farmers or farmer groups, i.e. support specialization among farmers.

Analysis of the local practices of production, storage and selection techniques, combined with a good knowledge of seed agronomy and technology can reveal important limitations to seed quality and opportunities for improvement. Availability aspects of seed saving will be dealt with in Section 2.5. Aspects that may be improved include:

o agronomy, harvesting time, and drying techniques that may affect germination, seed health (contamination by pests and diseases) or other quality aspects
o weed control, harvesting methods, and sieving/winnowing to solve seed purity problems
o seed contamination and practices to control the transmission of diseases
o seed selection methods to solve degeneration of the local varieties
o seed processing, storage practices and storage constructions.

Evaluating these points may be especially useful for on-farm seed production of local varieties. There are normally no formal seed sources (national seed programme or commercial enterprises) which do supply good quality seed of these local materials. Local varieties can also be considerably genetically degenerated through seed mixture, genetic drift or inbreeding. In such situations, seed selection practices can significantly improve the genetic quality of local seed and its availability.

The technical aspects of good seed production practices will be dealt with in more detail in Part B.

Participatory technology development
Suggestions for improvement can be tested in collaboration with selected farmers. Their comments and advice should be carefully analysed before presenting a technology for general extension. The definition and selection of the farmers-collaborators for this purpose is important (see Box 11.3). Key farmers may play an active role in the promotion of the improved practices, just as they are used in farmer-trainer or farmer-to-farmer approaches. In some cases, effective improved technologies can be extended to all farmers in the region through the existing extension organizations. Not all farmers may have the same conditions or priorities. Farmers will then either accept, reject or adapt the new technology based on their specific conditions and expectations.

An example of the adaptation of outside technologies is to add small quantities of Actellic (insecticide) when bean seeds are mixed with ash before storage in order to better control insect (Bruchid) damage. Another example is the use of a tarpaulin when sun-drying seeds where sudden rain showers can be expected during the dry season. This has the advantage, when seed is dried on special (concrete) drying floors, of reducing the contamination of the batch by different varieties and inert matter, and allows for the seed to be quickly covered in case of a rain shower. Well-cleaned fertilizer bags or other locally available materials are cost-effective for seed storage (although care should be taken to prevent seeds from coming into contact with left-over fertilizer).

The economic analysis of various options has to be part of the exercise. Such analysis should be done in a participatory way, and should not be left to economists only. Economists can calculate impressive returns on investment, but when the initial investment is beyond the scope of the farmers, or when the time needed for these positive returns is too great, the technology will not be accepted.

Community action
Even though saving seed is basically a practice of individual farmers, actions to improve seed production may be more effective when done in co-operation or at least in consultation with the community, involving both formal and informal community leaders (see Box 2.5). Involving communities increases the effectiveness of participatory technology development. Furthermore, the community is very important for seed security and has a function in the spread of technologies, information, and materials through farmer-to-farmer contact and exchange.

2.3.2 Supporting specialization in seed production
Co-operation with key farmer-seed specialists in the testing or development of new technologies can lead to specialization in seed production. Farmers who apply new technologies may become known as good sources of seed for other farmers in the community. They may have laid the basis for this reputation already before the intervention: collaborators in seed-related actions are often already the 'seed specialists' in the community.

One important action to improve seed quality and seed security in general may be to support these persons or farmer-groups in developing small seed enterprises. Small seed enterprises may be able to afford a better care to the seed crop in the field or slightly more expensive technologies such as small-scale threshers or seed cleaners. Seed enterprise development has become a trend in international co-operation after the poor performance of various national seed projects. Seed enterprise development can be large scale, i.e. developing an interest with existing businesses or large co-operatives in seed production; or small scale, i.e. encouraging farmers or farmer groups specializing in seed production for the food crops they generally produce.

Small seed enterprise development involves much more than just advising advanced farmers to sell seed instead of (food) grain. It involves the following steps:

○ analysis of demand and competition;
○ analysis of the economies of specialized seed production;

48

○ examining the logistics of processing and distribution;
○ planning for marketing;
○ the analysis of legal constraints; and
○ the actual start of the commercial operation.

This book does not provide a detailed action plan for seed enterprise development, it only presents some general points that need consideration when considering small seed enterprise development. These points are covered in more detail in Chapter 9.

2.4 Variety diffusion and seed exchange

2.4.1 Farmers' existing seed sources

Small-scale farmers usually prefer to use their own seed (see Table 2.1). It is the cheapest, most readily available, and of a variety that the farmer is familiar with. The farmer knows the seed quality, and the seed is available at planting time.

Table 2.1: Seed source in 1991 of maize and beans plots (% of total number of cases) in Costa Rica, Nicaragua and Honduras

	Maize			Common Bean		
	Costa Rica	Hon-duras	Nica-ragua	Costa Rica	Hon-duras	Nica-ragua
Farmers' own seed	79	75	81	58	79	72
Other local sources	19	13	12	21	15	14
Formal sources	2	13	6	21	6	13

Source: Wierema et al., 1993.

There are different reasons, however, why a farmer may be using seed from other sources:

o *To get seed of a new variety.*
o *The farmer was not able to save seed.* For example, because last year's harvest was too small and all the grain was eaten, because insects or moulds attacked the stored seed or because all the harvest had to be sold to meet sudden expenses.
o *To replace the farmers' own diseased or 'degenerated' seed.* Two different types of seed degeneration can be distinguished: genetic degeneration and a gradual reduction of the sanitary quality, commonly due to a build-up of virus in the seed.
o *Seed production conditions are not favourable.* Seed production may be difficult because seed-transmitted diseases cannot be avoided (e.g. for beans and potatoes under warm conditions).
o *Seed cannot be produced by the farmer.* For some crops it is just not possible to produce good seeds under local conditions. The production of hybrid seeds of maize, sorghum and vegetables requires the main-tenance of parental lines and a crossing technology which is too compli-cated for most local situations. Seed production of some biennial vegetables is not possible under local conditions because they require a cool period or specific day-length conditions to flower or to set seed.
o *The storage period from one harvest until the next planting is too long for seed quality to be maintained.*

50

o *The specialization required of the farmer* in producing for the market (seed free of stones, weed seeds, etc.) may result in a situation where seed production does not fit the level of mechanization and productivity of the farm any more.

These factors determine the demand for seed from a household, a community or a village. The demand for seed varies between crops. The fluctuation of the demand from season to season usually follows a pattern determined by the incidence of pests and diseases and the general yield level in a region (see also Section 9.2).

When the farmer is not using his or her own seed, there are different sources from which to obtain seed. The reasons for using seed from other sources commonly depend on quality and price:

o *Seed from a relative, friend or neighbour* can be a good option because it is of known quality. Because the farmer has seen the crop in the field during the previous season, both the variety and seed quality are known.
o *Seed from the market or from a middleman* can be risky. Seed from this source is often grain produced for consumption, sometimes with some selection for size or uniformity. Buying seed at the market or from a salesman is often a last option to obtain planting material.
o *National seed programmes* may be good sources for seeds of new varieties. National seed programmes may also have the mandate to provide certified seed for already recommended and widely used varieties. They usually do so for crops in which the commercial seed sector is not very active. These tend to be the crops in which the seed sector cannot earn much money, such as the self-pollinating cereals. Farmers do not have to buy seed of those crops since they are quite capable of producing seed themselves. Once they have the variety, they can save their own seed.
o *Commercial seed enterprises* have to make profits. For that reason, they will concentrate on selling seeds of crops and varieties that need regular purchases by the farmers. Thus, they prefer to specialize in hybrids and the seed of crops that are difficult to produce locally.

In most developing countries, the area planted with seed from the formal seed sector, i.e. from national seed programmes and seed companies, varies from 5 to 30 per cent depending on the crop (see Table 2.2). The importance of seed from the formal sector depends on the climatic conditions, the crop, the type of varieties being used and the success of new varieties and the quality of available seed. Rice, wheat and barley are self-fertilizing crops and farmers can relatively easily produce and store a good-quality seed. Consequently, the importance of the formal sector in these crops is small, farmers' seed being planted on 90–100 per cent of the crop area. For other self-fertilizing crops, such as beans and pulses, it is more difficult for farmers to produce disease-free seed and to store the seed over a long, warm and humid period. The formal seed sector may be more important for these crops; the area planted with farmers' seed is usually 70–80 per cent or more. In the (partly) cross-pollinated crops (maize, sorghum) and self-pollinated crops where hybrid seed is successful (tomato, other vegetables), the situation is different. The importance of seed from the formal sector in these crops can be considerable.

Table 2.2: Use of certified seed as a percentage of total area sown in a selection of African countries, 1985

	Maize	Sorghum	Wheat	Rice	Common Bean	Ground-nut
Angola	15	0	50	0	–	0
Botswana	66	100	–	–	0	<1
Lesotho	75	5	38	–	4	–
Malawi	10	5	19	12	4	0
Mozambique	10	5	13	0	–	–
Swaziland	98	21	80	100	2	0
Tanzania	14	9	15	1	<1	0
Zambia	70	0	97	0	12	<1
Zimbabwe	83	25	97	0	0	<1

Source: (figures on cereals: main report 1A, beans and groundnut figures from individual country reports, DANAGRO, 1988).

The area planted to seed from the formal seed sector is not the same thing as the area planted to modern varieties. The introduction of a new variety requires relatively large amounts of seeds to meet the demands of farmers who want to try the new variety, but once enough farmers grow the new variety, the maintenance and further diffusion of the variety can take place through on-farm production and seed exchange between farmers.

Even though the local seed-exchange mechanisms may be very effective, these depend heavily on traditions and are limited by natural or social barriers. This may create situations in which access to seed is not equally distributed among the different members and households of a farming community.

2.4.2 The strategy for strengthening traditional seed exchange
Activities that strengthen traditional seed exchange mechanisms and channels for the diffusion of seed and new varieties have a twofold effect: they improve varieties and seeds in the local systems, and increase the impact of the formal seed programme.

More directly, such activities aim to introduce new varieties or improved seed into the traditional seed exchange network. Such introductions may activate the mechanisms of seed exchange channels and lead to the diffusion of materials. Where active traditional seed exchange mechanisms already exist, it may be most effective 'to tap into these systems' for a rapid diffusion of improved materials. In order to achieve this, strategic amounts of seeds of new varieties or of improved seed quality could be introduced directly to crucial points in the local seed system network or via NGOs. It is therefore important to understand which individual farmers in a community, and which communities in a particular area or region, play a crucial role in the supply of seeds to other farmers. Access to seed through these traditional channels is not always equitable for all social and ethnic groups in the community (see Box 2.6). It is always necessary to confirm whether the better-off farmers, and the localities or areas with favourable seed production or seed storage conditions are to be the focal points for this type of activities. The significance of geographical barriers and possible socially separated groups in communities requires careful planning of the focal points for introduction.

Box 2.6 Seed and information exchange

The importance of these informal networks of seed distribution and information can be illustrated by the pearl millet variety, Okashana, which some farmers received for on-farm trials in Tsholotsho District, Zimbabwe. One farmer obtained Okashana three seasons ago from an extensionist. In a group discussion to which poor farmers were particularly invited, those farmers who lived in the same village but in a different *kraal*, said they had never heard of this promising variety. However, an informal communication network in which the extensionists play a critical role, ensured that farmers in other wards did know about it. The farmers of this network were mainly the better-off ones. The interested farmers obtained seed of this variety through the extension worker or directly from the farmer's homestead. The farmer and his wife were also asked to host a field-day to show farmers from other wards around and to spread information about the variety. The distribution of the seed was also facilitated by the farmer's relatives who were selling the seeds on his behalf in other wards. Good farmers can thus be important seed sources. Notwithstanding, it seems that poor farmers, even those who live within the same village, may not hear about a new variety. (Mheen-Sluijer, 1996; Commutec, 1996).

In poor conditions of food shortage coupled with land scarcity Ferguson and Mkandawire (1993) report that the poorest households tend to be women-headed households. The need to feed the family does not allow these women to save seed for the next season. They depend for their seed on bean purchases from the market, prior to planting. It has also been observed in several studies that poor households do not easily have access to the local seed exchange network, because they are not part of the social network of those farmers who usually have a surplus of produce that can be used for seed. In another case, however, a poor woman in the community was the recognized custodian of diversity, having great knowledge of varieties and seed selection.

For the introduction of these crucial amounts of seeds, demonstration plots, seed kits, seed fairs, and seed production with local groups or individual key-diffusers can be organized.

2.4.3 Demonstration plots

Demonstration plots in strategically situated farmers' fields and managed by the collaborating farmers themselves are effective in introducing a new variety or demonstrating improved seed quality: the sight of an impressive-looking crop capture the interest of neighbours and others passing by. Such demonstration plots can include only a limited number of new varieties or treatments, since the size of plots of the variety or treatment being evaluated should be relatively large in order to be visible to the passer-by.

An experimental plot can also be arranged on a farm in the community or on a nearby experimental station to which farmers are invited on one or two occasions. Transportation for the visitors may be required.

Implementing demonstration plots in school gardens, with women's groups or others, can also have a great impact. While the children or adults involved may take home the seed of preferred new varieties for next season's planting, they also gain confidence in handling certain treatments or in the use of seed of improved quality. Moreover, special days to visit the plots may be organized in co-operation with the school or the women's group.

A visit at around flowering time and another visit at harvest time is probably most timely in giving farmers an opportunity to appraise the character of the different varieties, or differences between treatments. To invite farmers to participate in the harvesting and to allow farmers to take home some harvested materials will increase the interest of the farmers in participating in a field day, and will also contribute to the introduction of those materials into the local system.

Demonstration plots may be set up like a conventional experiment, with or without adaptations to facilitate evaluation and comparison by visitors. The budget, technical skills and available time of the technicians and farmers implementing the demonstration plots are important considerations in the definition of demonstration trials. The planning of demonstration plots includes decisions about the number of visits, the number of farmers to be invited, which farmers to invite (entire community or individual farmers), communities from which farmers are to be invited, planning of transportation, and so on.

2.4.4 Seed kits

Distributing small packages of seed has been shown to be effective in diffusing materials. These kits contain one or two samples of seed together with inputs and information on how to use the kit. In most cases these kits are intended to be only a part of a larger technology package. Farmers can evaluate the material in their own way and save most of the harvest as seed for the next planting. Seed kits may be particularly effective in areas that are poorly covered by the formal sector and where close assistance and follow-up are difficult as a result of remoteness or inaccessibility. Geographical inaccessibility may, however, also make the even distribution of seed kits difficult to achieve. A follow-up study after two to four seasons will show how many farmers are still growing the variety and how much exchange between farmers has taken place. Such follow-up also serves to elicit information on the performance of the distributed materials and farmers' opinions.

The packages may be distributed to key farmers, or offered for reasonable prices at local fairs, on field-days or at seed fairs. The amount of seed in the packages should not be too small, as this may result in farmers' losing the crop in the field and not being able to save seed. Smaller packages also mean that instead of one season, the farmer may need two or more seasons before enough seed is produced to share, or to exchange with others. A project in Nepal offered farmers 1 or 2 varieties, to plant 0.05 ha rice (see Cromwell, 1990), a project in Zambia distributed 400 samples of beans, each to be planted on 0.03 ha (Grisley and Shamambo, 1993). The distribution of seeds from cross-fertilizing crops requires larger samples than those of self-fertilizing crops and vegetatively propagated crops. If a sample of a cross-fertilizing crop is too small the result may be inbreeding when a farmer multiplies the seeds over a number of years. For cross-pollinators, a rule of thumb could be: one tenth of the amount of seed a farmer normally uses for planting a plot.

2.4.5 Seed fairs
Regional fairs after harvest are traditional yearly events in the Andean Region of South America where farmers exchange produce and seed. In Peru, NGOs in collaboration with staff members of universities and the national research institute INIAA used these traditional events to organize 'seed fairs' with 'diversity contests'. Two to three months before the fairs, letters of invitation were sent to the community leaders, indicating the objectives, date, place and rules for the crop diversity competitions. During the fair, each farmer or farmer group was registered and asked about: (i) the species presented, (ii) name of the varieties or landraces, (iii) characteristics of the material, and (iv) agronomic information. The juries were composed of two technicians and two farmer-experts. Farmers with the greatest number of varieties or largest ears or fruits received prizes in the form of agricultural tools, sacks of seed or certificates. Similar fairs have been organized by NGOs in other parts of the world (for example by ITDG in Zimbabwe and Kenya), and have proved to be very successful in making farmers aware of diversity in crops and varieties and in encouraging farmers to think about the importance of diversity. The principal purpose

was to stimulate the maintenance and use of crop genetic diversity, but the diversity contests have proved to be effective in bringing out materials, stimulating both the exchange of seeds and information.

2.4.6 Seed production with key producers

Apart from providing samples of seed from new varieties to crucial farmers or communities in the local seed network (see introduction to this chapter), technical support for these farmers is needed to improve their seed production and thereby improve the availability of good-quality seed in the traditional seed channels. Activities may address the technical aspects of seed production (agronomic production practices in the field, seed selection practices or the storage of seed) as well as the more organizational aspects of setting up collective seed production.

Working with community or farmer organizations is an obvious choice when such groups already exist, such as in the case of a number of NGOs in Latin America. The Nicaraguan NGO CIPRES assisted farmers in producing seed of improved bean varieties. Recently the Ministry of Agriculture bought seed from the farmers for national distribution. Another example is from Ecuador, where an NGO supported organized farmers in potato seed production (Box 2.5).

2.5 Support to local seed security

Seed security can be defined as 'the sustained ability of all farmers to have sufficient quantities of the desired types of seed at the right time'. It has two aspects: the availability of and the access to quality seed. It does not only refer to the quantities and qualities of seed, but also to the timing (i.e. the availability of seed at planting time), the finance (ability to have or purchase), and equity (access to available seed for all farmers in the community). Seed security has to be addressed at different levels: household, community and national. The best general approach to increase seed security is to strengthen the local seed system. General poverty alleviation and consistent national seed policies are also necessary. The following section refers principally to activities to improve seed security through the availability of seed at the household and community level.

2.5.1 Seed banking

Community seed banks can permanently strengthen local seed security by providing an easily accessible local outlet for seed, at a reasonable price and quality. Depending on the organization, it may have the added advantage of encouraging and contributing to local enterprise. The principle of a seed bank is that a safe place is established in which to store seed. Households commit seed to it at harvest time, and may take it out again for planting next season. The objectives and organization of a seed bank can vary. Seed banking can aim at improved storage conditions for the seed, at providing a seed reserve for a community or at maintaining a germ plasm collection which is easily accessible by farmers.

Communal storage places provide conditions for storage which are more favourable or more protected than in the traditional storage. Seeds are brought in at harvest time and taken out at planting. Individual farmers' seeds may be mixed or maintained separately. This is an option in situations where the implementation of improved storage facilities is cheaper on a community basis than on an individual basis.

A *seed reserve* can be a valuable resource to which community members turn in emergencies, when left without seed. Seed taken out may have to be paid for or replaced after harvest. This requires clear community rules about acceptable levels of withdrawal and the quality of contributions. This is an option that will contribute to seed security in situations of high risk and when other seed exchange systems prove inadequate. If the seed is being kept as a reserve, it may not be needed when disasters do *not* occur. The volume of seed being stored must therefore be a function of the seed needed for security, the storability of the seed and the costs of keeping the seed.

If a price has to be paid for the seed from a seed bank, there needs to be agreement in advance how any funds generated from the sale of seed will be used: will they be used exclusively for seed-related activities, or can they be used to fund other village development activities? Also, what happens if the harvest fails and seed bank stocks cannot be replenished?

In Northern Ethiopia, shortage of seed is dealt with through community seed banks (see Box 2.7). Similar approaches are now being tried in other regions in sub-Saharan Africa, where droughts and disasters have been occurring frequently. There is not yet enough evidence on how community seed banks function in communities that do not have a tradition of operating such structures, but the interest in many communities for setting up such a system does show that the need for improved seed security is very strong.

A village *gene bank* conserves seeds of local varieties that are under threat. To operate successfully, specialist knowledge is required to ensure seed is selected properly in the field, and grown out properly on a regular basis to maintain the seed quality. In the case of genetically heterogeneous landraces, an adequate sample size is required.

There are many ways to administer a seed bank: in some cases agencies donate the initial seed stock; while in others seed is saved by local households. Sometimes the seed bank is operated by a village committee, elsewhere it is controlled by agency personnel. In some cases seed of a single crop is stored; in others seed of many crops and varieties are stored. The store may be constructed by the community, or an existing building may be used, or else the agencies involved may build the store to technical specifications suitable for seed storage.

Box 2.7 Traditional seed banking in Ethiopia

In Tigray the idea of establishing community seed banks originated from discussions after the famine of 1984–85. People had noticed that some farmers achieved a better harvest than others in spite of equal crop husbandry, uniform climatic conditions and similar soils. They believed that the differences could be explained by the quality of the seeds, since those who got the better harvest were known to be particularly competent seed selectors. In the same discussions people also complained of the exploitation of drought victims by seed lenders who charged exorbitant interest. These points were made independently in meetings at several places in Tigray. REST (Relief Society of Tigray) heard these ideas and complaints and proposed the establishment of community seed banks through which the benefits of traditional seed selection could be extended and through which non-exploitative credit for seeds could be provided.

Various NGOs made funds available for the initial purchase of selected seeds for the seed banks and activities started in 1988. By 1991 banks were operational in 42 *Woredas* (sub-districts) covering a major part of Tigray. These banks depend on the locally available plant genetic resources and the indigenous knowledge of seed selection.

The seeds banks are owned and managed by the '*baito*', an elected body at the *Woreda* level. The *baito* identifies the experts in traditional seed selection and good seeds are purchased on their advice. The selected seeds are entrusted to female seed keepers. Loans are granted by the *baito* and borrowers receive the seeds at planting time for the price which the *baito* paid when it was purchased. The borrowers have to pay back in cash 6–9 per cent interest after harvest. In cases of crop failure, the *baito* accepts a one-year delay of repayment without charging additional interest. There seems to be strong social pressure for repayment and so far most of the seed banks have managed to maintain their initial capital fairly well. Seed selectors and seed keepers provide their services free of charge. The administrative tasks are shouldered by the *baito*.

Source: Trygve Berg, 1992.

Farmers may need training in seed technology, especially selection and storage. Care is needed to make sure that the committees set up to manage the seed bank operate equitably and accountably. Care is also needed over the initial choice of technology (for seed bank construction, and equipment for drying, cleaning, storing and treating the seed) so that it can be maintained easily by the local community.

A number of manuals exist giving useful guidance on seed banking, including one produced by RAFI (RAFI, 1986), and the Seed Savers Network (Australia)/UK Heritage Seed Programme (Cherfas and Fanton, 1996), providing community seed bank training to groups in developing countries.

2.5.2 Seed from relief programmes

In most circumstances, seed relief should be a short-term intervention. It is rarely appropriate or feasible while an emergency is still at an acute phase: it usually starts during the settling-down period. If agencies want to carry on with seed activities after the first few growth cycles following an emergency, then they should move on from seed relief to longer-term seed capacity-building. Repeated distributions of seed relief after the first few post-emergency seasons is not appropriate: it interferes with restoring the local economy and re-establishing local seed supply.

The underlying rationale for seed relief is that it helps to re-establish a 'self-help' mode within communities affected by emergencies: once families have seed and basic tools, their dependence on external sources for their livelihoods is reduced. It is a waste of resources getting involved in seed relief, however, unless there is a clear indication that a lack of seed is the key factor preventing families from returning to 'self-help' mode. Even after severe droughts or armed conflicts, seed is often still available within communities (from secret stores, or via traditional seed supply lines) and other items – such as drugs, tools and building materials – may be in much greater demand. The free distribution of seed in such a situation may be very damaging to the restoration of the local seed system and the use of locally adapted genetic material. In these circumstances it may be more useful to provide food aid, so that families are not forced to eat their hoarded seed. Diagnosing and planning emergency seed provision is complex, not least because of the need for acting fast.

Box 2.8 Seed provision in an emergency

The key components of seed provision during and after emergencies are:

○ pre-planning to assess whether or not seed is needed and/or is with;
v○ identifying the type of seed to work with;
○ deciding which agencies and structures to work through and/or with;
○ identifying the type of seed to work with;
○ selecting an appropriate source of seed;
○ identifying which supporting services should be provided together with the seed (e.g. fertilizer, tools, etc.);
○ identifying target recipients for seed;
○ calculating the quantity of seed needed;
○ organizing the logistics of seed distribution;
○ tracking (monitoring) seed;
○ evaluating the impact of seed; and
○ deciding to stop.

Seed provision during and after emergencies, ODI, 1996.

Where seed relief *is* appropriate, the aim should be to distribute seed that is as close a possible to what the community was using before the emergency: not just seed of the same crops, but also of the same *varieties* (see Box 2.8).

The aftermath of an emergency is not an appropriate time to experiment. As a recent FAO report stated:

food aid, combined with the importation of often poorly adapted seed varieties, can lower yields and keep them low for years. Whilst addressing the immediate crisis, such practices can exacerbate hunger, undermine food security, and increase the costs of donor assistance well into the future (FAO, Global plan of action (1996) p.16)

3. Biodiversity and *in situ* conservation

3.1 Biodiversity

Over the last few decades we have begun to realize that we are losing the wealth of living forms on Earth. The consequences of this loss cannot be foreseen, but it is generally recognized to be one of the greatest problems we face for the future. This diversity is often seen as having developed over the history of the Earth. In agriculture, this diversity is also shaped by the farmer's guiding hand, using and further developing the biological resources at his or her disposal. In these natural and human processes, biodiversity has developed with a continuous gain and loss, though it is clear that in recent decades the balance has become negative.

'Biodiversity' or 'biological diversity' is a relatively new term which emerged about 15 years ago. Biodiversity refers to the variety of life forms, the genetic diversity they contain, and the assemblages they form (see also the Glossary). Biodiversity at any one level is the sum and the product of different lower levels of biodiversity. Generally, three levels of biodiversity are distinguished:

○ ecosystems diversity;
○ species diversity;
○ genetic diversity within species.

Biological diversity is found in forests, savannahs, grasslands, deserts, lakes and even in towns (e.g. household gardens, parks, etc).

Genetic diversity is also a relatively new term, and refers to 'the variation within and the variation between populations of species'. It can be measured in terms of, for example, variation between genes or between DNA or amino acid sequences. It can also be measured in the number of breeds, strains and distinct populations (Heywood and Watson, 1995). Some use the term 'genetic variation' within species only, others for within as well as between species.

Species diversity normally refers to diversity among species. A single species includes organisms that are closely related, morphologically one distinct unit and, when mating, form reproductive offspring. Species diversity, therefore, refers to the diversity of species in a forest (e.g. insect species, bird species, plant species), different fungi in the soil, fish species in the river, and plant species in the garden.

Ecosystem diversity is both the sum and product of the other two levels of diversity. Different organisms have different roles and functions. Different organisms also interact and thereby form a community that has developed or evolved in a physical (dry, cold, wet, hot, fragile, poor or rich) environment. The diversity of roles, functions and interactions within the particular environment represents the ecosystem diversity. The scale of ecosystem diversity is important, as one square metre in a forest can be considered an ecosystem, while the entire Amazon or Pacific may be considered one as well. Ecosystem is a hierarchical term. It considers the one square metre in the forest a part (or subsystem) of a larger system, the Amazon. The Amazon in turn is a subsystem of the ecosystem that we call the Earth.

Since biodiversity is composed of different levels of diversity, it is not possible to express it in a single quantitative measure. The number of alleles, number of crop varieties, the number of plant species, the number of strains of fungus, the number of ecosystems, are different aspects of biodiversity, and they are difficult to sum. Furthermore, the frequency with which particular alleles, varieties or insect species occur is also important, as well as their distribution in space and the fluctuation of their occurrence over time.

3.2 Biodiversity in agriculture

The three levels can also be used to describe biological diversity in agriculture, in a similar way:

○ farming systems or agro-ecosystems diversity;
○ crop, animal and other species diversity; and
○ varietal and other genetic diversity.

Genetic diversity in agriculture can best be analysed at the level of varieties of crops and breeds of animals.

Species diversity in agriculture relates to the different species: the different types of crops, animals and fungi, and also the diversity of crops that can be found at one farm, or the total collection of cereal crops used to feed the Earth's people.

Just as for natural biodiversity, farming and agro-ecosystem diversity should be viewed at different scales. One farm can be treated as an agro-ecosystem, or an entire region can be considered. The diversity of one farm in the mountains of the Andean Region is very different from the level of diversity that is encountered in the savannah of Zimbabwe or the polder of the Netherlands. It is important to realize that the agro-ecosystem is the sum and product of the other two levels of diversity.

One element strongly distinguishes agro-biodiversity from natural biodiversity. Agriculture is the way humankind uses its natural biological and physical resources to feed itself, to cure disease, construct shelter, make clothing, and to earn income. The role of farmers in the development of diversity in agriculture is very important. Many different agro-ecosystems, crops and varieties can be found all over the world, and it is not only natural conditions that have contributed to this diversity: differences in the human population have contributed to this enormously. Some people therefore consider human diversity, with its social and cultural elements, to be a fourth level of diversity. Farmers' knowledge and practices on how to grow crops, and to understand their specific medicinal purposes, etc. are elements of this diversity.

3.3 The loss of biodiversity

The degradation of natural ecosystems throughout the world, but especially in tropical regions, has been well documented by scientists. For example, the lowland forests, such as the Amazon, were until recently the least-disturbed natural areas in the tropics. These forests contain more than half of the total species on Earth. They are now experiencing human exploitation at a very fast rate. Shifting cultivation is a form of agriculture which is well adjusted to the physical conditions of the humid tropical forests. The intensification of agriculture, population pressure, cattle ranching, and timber exploitation, however, are leading to an accelerated loss of diversity in this ecosystem.

The causes of the loss of biodiversity are biological, but the underlying origin of these problems lies in the social, economic and political processes that operate on a world scale. The forces at this global level are enormous, but the options open to individual people are limited. Actions on the part of farmers, consumers and people living both in the North and South can contribute to reversing the deterioration in their living environment, however, and this may benefit not only the world now, but also the world of their great grandchildren.

The loss of natural biodiversity
The diversity of living things never has been and never will be static. Biodiversity fluctuates in time as evolution adds new species, and extinction takes them away. Evolution and extinction are natural processes, they are the responses of populations of organisms to changes in their physical and biological environment. Change is, in a very real sense, a basic fact of life. Although change appears to be a norm, we are now concerned about the loss of biological diversity because the loss is due to the destruction of habitats, and because over-exploitation today has a different origin, order and magnitude to those recorded earlier.

The loss of biodiversity in agriculture
Similar processes to these in nature are eroding the biodiversity in agriculture. The loss of biodiversity in agriculture takes place at the three levels. Farming and agro-ecosystems change, as particular crops are no longer cultivated or are being marginalized. Most prominent in agriculture is the process we call genetic erosion or the loss of genetic diversity.

The process of replacing local, indigenous, traditional varieties or land-races by modern, high-yielding varieties is often equated to the loss of genetic variation, and consequently is known as genetic erosion. Agricultural processes must be examined with respect to the loss of genes or gene combinations. Gene replacement occurs when indigenous varieties are replaced by introduced ones. Genetic erosion can be seen in two forms: genic or allelic erosion and genomic erosion. The replacement of landraces by new ones within a crop results in the complete replacement of those alleles which differ between the local and the new variety. The replaced genes or allelic forms are lost or reduced to such low numbers that they can easily be lost if not conserved or used elsewhere. In addition, the specific combination of genes that occurs in the replaced variety is lost.

Apart from the physical loss of genes, gene combinations, genes or – more visible to farmers – varieties, knowledge of specific crops and varieties are threatened by a similar process of erosion. The modern development of agriculture leads to a globalization of agricultural practices, and specific skills of how to manage and use particular crops or varieties erode. Monica Opole from Kenya refers to women in the rural areas of her country who now send their daughters to school where they learn how to grow tomatoes and cabbages in order to be modern mothers and farmers. Their mothers have started to realize that their daughters are not taught the specifics of how to grow indigenous leafy vegetables any more. This knowledge of important plants which grow plentifully at their farms is gradually lost. It is not only the knowledge of the species but also the knowledge of their special medicinal and culinary properties, and ways to process and prepare them, that are lost.

Another form of genetic erosion occurs at the intermediate level between agriculture and nature. As described in Section 1.1 on centres of origin and evolution, most crops still have related wild and weedy species. Where crops are grown in the direct environment of these non-domesticated species, some introgression may occur. The maize–teosinte complex in Mexico is an example of a system where a wild relative grows in the neighbourhood of the crop. Various researchers have investigated this interaction, looking for specific characteristics of the local maize varieties which may originate from the wild teosinte plants. All plants related to a crop species may eventually be a source of important traits for crop improvement, through introgression, modern plant breeding and biotechnology. Owing to the destruction of specific habitats, such wild relatives of important crops may disappear, leading to erosion at both the crop species and genetic levels.

3.4 Conservation strategies

Based on the definition of biological diversity, biological conservation can be regarded as the effort to maintain the diversity of living organisms, their habitats and the interrelationship between organisms and their environment. Conservation cannot be just about individual plant and animal species, since these life forms and their functions cannot be maintained without maintaining the environment or ecosystem in which they live. Different conservation practices can best be divided into approaches focusing on the conservation of the habitat or ecosystem, and approaches focusing on the conservation of genetic diversity. In nature conservation, the focus is mostly on the conservation of natural habitats and ecosystems, while for agricultural biodiversity the focus has been primarily on conserving genetic diversity. There has been considerable development in these approaches over the last two decades, but only recently has there emerged interest in integrating these two approaches, and in the human component in conservation, which is particularly important in the case of the agro-biodiversity.

The conservation of agro-biodiversity
Conserving crop genetic resources, with the aim of maintaining genetic diversity, at first used the strategy of *ex situ* conservation. Over the last decade, *in situ* conservation has developed as a complementary strategy in which the maintenance of the agro-ecosystem plays an important role. The two strategies are defined by the location where the conservation activity takes place: *ex situ* conservation refers to the conservation of components of biological diversity outside their natural habitat, while *in situ* conservation refers to the maintenance and recovery of viable populations of species within their natural environment. In the case of domesticated and cultivated species, this means that these need to be maintained in the environments in which they developed their distinctive properties (definitions according the Convention on Biological Diversity; UNEP, 1992). Thus, to maintain agricultural diversity, the agricultural systems in which the diversity developed need to be maintained.

3.4.1 Ex situ *conservation*
Ex situ conservation of plant genetic resources is effected through genebanks, where samples of seeds or other plant material are stored under controlled conditions of temperature and humidity, mostly in refrigerators or deep freezers for medium (4° C) to long-term (–20° C) storage. The aim is to conserve as much as possible of existing genetic diversity. Many genebanks were created to provide breeders with readily available, relevant basic materials for breeding programmes. Now, the need to ensure the availability of genetic information for future generations is gaining importance. Materials are collected through plant exploration and are briefly discribed ('passport data') before being stored. The techniques of *ex situ* conservation are generally considered suitable for the conservation of crops, crop relatives and wild species.

The conservation of germplasm in field genebanks is another form of the *ex situ* strategy. It involves the collection of material in one location and the transfer and planting of the material in a second site, for example on a research station. It is the answer for *ex situ* collection of perennial species and root crops, such as rubber, coffee, bananas, cassava, sweet potato and yam. Seeds of such species cannot be dried and frozen without loss of viability. Alternative methods of *ex situ* conservation for these crops, such as cryopreservation, are rather complex.

Disadvantages of ex situ *conservation*
Over the past few decades, genebanks have proved vulnerable to several problems, including failing infrastructure (electricity cuts) and under-funding. In many genebanks the germination rate of accession falls well below the internationally agreed acceptable level of 85 per cent, (e.g. the famous genebank of the Vavilov Institute in Saint Petersburg). This may mean that in genebanks materials or genes are lost. Furthermore, there are very sad stories of wars and political instability threatening *ex situ* collections, such as, for example, the genebank of the Rice Research Station in Rokupr in Sierra Leone, which was destroyed by rebels.

An important characteristic of genebanks is that they 'freeze' the evolution of the stored materials. Because genotypes are taken from their original environment, they are no longer subject to the continuing changing environment and farmers' crop and seed practices. If properly stored and reproduced, genebank accessions can be maintained with little genetic change over a long period. Yet, if the same populations were allowed to survive *in situ* – on the farm in the area where they were collected – they might have evolved and adapted to changed production conditions and practices. In the farmers' fields, germplasm may co-evolve with diseases and pests, changing farming systems and climatic conditions, while material stored over a long period of time may theoretically not be adapted any-more if conditions in farmers' fields changed considerably.

The information on accessions in genebanks also represents a limitation of *ex situ* conservation. The quality of the material stored is not only dependent on its viability but also on the availability and accuracy of the information on the stored material. Passport data rarely include properties or uses described by farmers or describe the agro-ecological conditions of where the material originates. Plant explorers often spend only a few min-utes on each sample they collect. There is no time to chat to farmers and record their knowledge, and so the linkage between the farmers' know-ledge and the biological material is broken.

Genebanks have also been criticized in the global debate on access to genetic materials. For local communities, *ex situ* genetic resource collec-tions are effectively extinct. When placed in long-term storage facilities the material is practically unavailable for the farmers. Materials in the bank are usually made available to plant breeders and researchers, but not to the farmers and communities whence they came.

Another issue is related to the property rights of the materials stored in genebanks. According the Convention on Biological Diversity (UNEP, 1992), national states have the sovereign rights over biological resources. This may be in direct conflict with the interest of local communities, who may want to maintain their germplasm themselves or to keep it in custody in formal genebanks under a black box agreement. In a black box arrangement, the material is stored in the formal storage facility, while the documentation remains with the owner, and the material can be taken out only under their authority.

3.4.2 In situ *conservation*

In situ conservation aims to maintain organisms in their natural habitat, allowing evolution to continue and maintaining adaptation. A genebank conserves only samples from the genetic diversity in a system and thus may miss some relevant genes. In contrast, *in situ* conservation aims to maintain all genetic diversity by maintaining the entire system. This strategy has been adopted and adapted from nature conservation. *In situ* conservation is particularly concerned with conserving semi-wild species or the wild relatives of crop species. It is particularly relevant for habitats where crops and their wild relatives occur together, and which are under such pressure that the wild relatives might disappear.

These habitats may be natural, but they may also be strongly human influenced, such as, for example, the Biosphere Manantlan in Mexico, where farmers grow maize and where Tripsacum, the wild relative of maize, still occurs.

Other examples of human involvement in *in situ* conservation is the conservation of grasslands or pastures. *In situ* conservation implies that the grazing intensity of such pasture is maintained in such a way that certain populations of wild species are maintained. Stopping the grazing could change competition between species and result in the disappearance of the target species. A project of the Global Environment Fund (GEF) from UNDP on *in situ* conservation in Turkey aims to establish 'gene management reserves' in areas that are rich in crop-related wild species, such as wild wheats. Controlled grazing, mowing or fire management to discourage competing perennial species, especially perennial grasses, are part of the conservation work (Ertug Firat and Tan, 1997). Another example of *in situ* conservation is the national coffee conservation programme of the Biodiversity Institute in Ethiopia. A special effort is being made to conserve the semi-cultivated coffee of peasant farmers in areas where forest coffee occurs spontaneously. This *in situ* conservation complements to *ex situ* field collections being maintained by the genebank (Worede, 1997).

3.4.3 On-farm conservation

In situ conservation of plant genetic resources can also specifically target the conservation of local varieties or landraces. In this case, the farm or agro-ecosystem is considered to be the habitat where the genetic diversity developed and originates from. Many conservationists tend to consider on-farm conservation of local crop varieties only relevant in regions where the crop was domesticated or where it still occurs in combination with its wild relatives (the centre of origin). This is also suggested by the definition of *in situ* conservation by the Convention of Biological Diversity, given earlier. This is debatable, however, since evolution outside the centres of origin has resulted in different genetic constitution of the materials, by which it can be argued that these materials originate from the farms where they were further shaped and maintained. Furthermore, a widening of the interpretation of *in situ* conservation is brought about by recognizing the role *in situ* conservation can play in complementing *ex situ* conservation. This means that, for example, the diversity of maize in East Africa and Asia, or of common beans in the Great Lakes region of Africa make these areas relevant for on-farm conservation for maize and beans, respectively.

On-farm conservation involves the maintenance of crop varieties or cropping systems by farmers within local agricultural systems. Local varieties or landraces are still the dominant type of materials in many marginal production environments. Farmers sow and harvest these varieties each season, mostly using their own saved seeds. Thus the landrace is continuously grown in the specific production environment of the farmers. It is adapted to the local environment and is likely to contain locally adapted gene combinations (see Chapter 1). This local reproduction and related seed exchange represents an integrated form of seed production, crop development and conservation, all in the same process and all carried out by farmers. From the conservation perspective, it is a local and dynamic plant genetic conservation system. However, it is not the farmers' objective to conserve the genetic diversity; the farmer aims at sustaining or improving his or her family's livelihood. In sustaining and improving the livelihood, farmers' access to genetic diversity plays an important role. It is here that the agenda of the conservationists meets the agenda of development specialists. For on-farm conservation to be effective, it has to strengthen this production system and coincide with the farmers' objectives. Or, in other words, for the farmer to be able to maintain genetic diversity on farm, the farming system needs to be strengthened and developed in the first place.

Since it is not possible to oblige farmers to continue to grow local landraces, on-farm conservation can be achieved only if conditions continue to be attractive for farmers to use genetic diversity. Of course, it is possible to compensate a farmer for continuing to plant a particular local variety, but this would be difficult to organize on a large scale and would immediately stop when financial means for compensation were exhausted. A more sustainable form of on-farm maintenance of genetic diversity is to stimulate continued use of landraces by farmers. Maintenance is the logical result of continued use. Strengthening this local system in which conservation is integrated with seed production and crop development is the best strategy to facilitate continued use. Improving the quality of seed of the local varieties may make them more competitive or attractive in comparison to improved varieties. Thus, seen through the eyes of the conservationist, support to local crop development and seed production are important tools in on-farm conservation. Futhermore, other factors such as marketing and legislation all, in one way or another, are related to the local system and can potentially contribute to the use of diversity by farmers, and thus support on-farm conservation.

Linkages between on-farm use and the maintenance of agro-biodiversity
Support to the local system is described in great detail in Chapter 2. Improving the quality of seed, helping the diffusion of new and local varieties and stimulating seed exchange, and participatory plant breeding all contribute to the local seed system, and may play an important role in on-farm conservation, as long as utilization of genetic diversity is kept in focus. It should, however, be realized that such incentives can also lead to variety displacement or genetic erosion. This is inherent in the dynamic nature of the local seed system. Looking at on-farm conservation in this way, there exists a range of breeding and seed-related activities aiming at agricultural development and providing access to genetic diversity (see Box 3.1) that can be very direct contributions to on-farm conservation.

On-farm conservation activities – working together with conservationists
The fact that there are a range of crop improvement and seed-related activities that are effective in supporting on-farm conservation of genetic diversity means that conservation and development-oriented people need to collaborate. Box 3.1 elaborates a number of options to link the local system with institutional conservation, and vice versa. These activities can range from primarily a monitoring of changes in the use of genetic diversity in time and space, to direct interventions or collaborative development-oriented actions.

Enhanced support of the formal system to strengthen the local system can contribute considerably to the conservation and utilization of biodiversity in agriculture at the farm level. Taking the local seed system as a starting point for seed production, as in this book, is therefore at the same time a starting point for the implementation of on-farm conservation of agro

Box 3.1 Linking institutional conservation to local seed systems to contribute to on-farm conservation

○ The restoration and re-introduction of PGR (plant genetic resources) to rural communities aims to increase access of farmers to both their indigenous and exotic germplasm. The Ethiopian Biodiversity Institute is involved in such activities, restoring indigenous germplasm to their original communities (Worede and Mekbib, 1993).

○ Supporting and strengthening existing local seed bankers (Benzing, 1989) or those farmers or communities that have a traditional role in maintaining a higher degree of diversity in their community.

○ The establishment of community seed banks (Berg, 1992; Magnifico, 1996) to maintain diversity at the rural community level. In such activities, the initiative comes mostly from development and/or conservation organizations. The linkage with seed multiplication in the institutional system is apparent. Seed banking activities by NGOs are mostly based on either a seed security, or a strong local empowerment and autonomy perspective.

○ Landrace enhancement is another example of a conservation action with a strong association with variety improvement (Worede and Mekbib, 1993). In landrace enhancement, it could nevertheless be asked who controls the enhancement: the conservationist or farmer? And with what goals? Crop improvement, continued adaptation or conservation? If the seed system is taken as the basis, the farmer should be a crucial decision-making player in this activity.

○ Documentation of indigenous knowledge is an area that requires special attention as it will enhance the quality of genebank collections. Methods for data collection, processing and the management of indigenous knowledge associated with PGR are barely developed. This area broadens the field of PGR management across the boundaries of the biological sciences into the social sciences.

○ Geneticists and PGR-specialists have an important role in monitoring the dynamics of local management of PGR and assessing the impact of interventions of the institutional system on the local system. As farmers may adopt other varieties, stimulated by market forces or government policies, or as diseases may wipe out certain landraces, immediate interventions to collect materials may prove necessary. This specific 'watch dog' role puts PGR-specialists in a similar role to conservation biologists monitoring ecological processes *in situ*. The focus in management on-farm are the agro-ecological processes of local crop development.

Source: W. de Boef, KIT.

biodiversity. On-farm conservation is more than a single conservation action, it addresses all elements of biodiversity in the agro-ecosystem at the same time, at the species, genetic and human level. It is also clear that for on-farm conservation to work, close collaboration between farmers and other actors in plant genetic resource management (plant breeders, seeds people, development workers, extensionists) is essential. Such an approach will also result in integrating conservation aspects into other interventions addressing development. This may be not only a way to conserve agro-biodiversity, but also help to use genetic diversity for a more sustainable agriculture.

4. Support from the formal sector

Each actor in the seed sector, ranging from the local farmer-producer to the multinational seed company, has a particular expertise and a useful role to play in ensuring secure seed supply, whether at the local or national level. Existing local seed systems should therefore not be looked upon in isolation from the formal sector, including the organizations producing certified seed, the national farmer support services and banking, and the policy levels. This chapter therefore investigates the role of these three parts of the formal sector in relation to enhancing seed security at the community level.

4.1 Contributions by the formal seed sector

The formal seed sector produces (certified) seed within an institutional framework, involving scientific plant breeding, structured seed production through generation systems, systematic seed-quality control, and subsidized distribution (public seed sector) or commercial marketing (commercial seed sector). Contributions to the development of the local seed sector can be made at every level of the seed chain.

Scientific plant breeding, even though it may concentrate on producing varieties for high-input commercial farming, can produce varieties that are useful for less-endowed farmers. The testing of such varieties under local conditions can contribute significantly to the local seed systems. Where (public) plant breeding can be geared specifically to more marginal and remote conditions, communities may benefit even more when they can participate in the plant breeding and variety evaluation process. Section 3.2 and Chapter 12 elaborate on this. Furthermore, scientific plant breeding can identify genes and can make valuable materials available through pre-breeding.

Formal seed production has generated significant progress in understanding seed-quality factors, and in developing solutions to such problems. A large part of section B is devoted to this aspect. Adapting such techniques to local conditions or improving local methods through the increased understanding of seed physiology and seed pathology can contribute significantly to local seed quality and seed security.

Seed quality-control systems are well developed in many formal (national) seed systems. Seed laboratories, field inspection and post-control facilities are available in many countries. Very often, these seed certification services (or 'authorities') act as a kind of police force in the seed market. This is useful in some cases, but the balance between policing and assisting seed quality is sometimes lost. Local seed support projects often have a great need to measure seed quality, for example when local seed drying and storage methods are compared. Secondly, emergency seed provision activities need to assess the quality of the seed that is distributed. The active involvement of formal seed quality-control agencies in local seed system support activities is therefore extremely useful, but in practice this opportunity is not (yet) taken up. Especially in small seed-enterprise development activities, the guidance (and not so much the control) of these formal institutions is necessary.

Formal seed marketing and distribution channels are often too specialized to support local initiatives effectively. The formal seed sector has, however, generated an understanding of seed enterprise development, which can also be used for sustainable community-level seed provision initiatives. Section 2.3.2 and Chapter 9 deal with this aspect. Especially in advanced stages of local seed enterprise development, experiences in the field of planning, logistics, promotion and selling can be very useful.

4.2 Legislation

4.2.1 Different regulations
In most countries the production and distribution of seed is regulated by one or more of the following laws:

○ seed law;
○ phytosanitary law; and
○ plant variety protection law.

Such laws are primarily made to regulate certified seed production and seed importation. They are not always intended to regulate local seed supply, but the wording is often such that small-scale seed production does fall within the scope of these laws. For any farmer, co-operative, commercial or development organization, it is necessary to be aware of the possible implications of these laws in order to avoid possible disputes with authorities.

4.2.2 Seed laws
There are many differences between seed regulatory systems in different countries. The main features of national seed laws are, however, similar. Seed laws normally regulate the official release procedures for varieties, the arrangements for seed quality control, and the organizational framework for seed sector development.

Variety release
Varieties have to undergo a registration procedure in most countries. Where plant breeding is done primarily in public institutions (research centres and

universities) breeders have to provide data from variety trials, and give evidence that the variety is adapted to the country's agro-ecological conditions. They may also have to demonstrate that the product quality is acceptable for local consumption or export. Such evidence is normally derived from a prescribed number of variety trials throughout the country, the so-called V.C.U. trials (multilocation trials for Value for Cultivation and Use). Yield levels and disease resistance are commonly the main criteria for acceptance of a new variety. Other characteristics may be taken into account as well, such as seed colour (cowpea), straw yield (sorghum), resistance to lodging (maize), cooking time (beans), aptitude to shattering (soya bean). Furthermore, the breeder has to describe the variety in such a way that it can be identified. The standardized procedure is called a D.U.S. trial (trial for Distinctness, Uniformity and Stability). The results from both V.C.U. and D.U.S. trials is then reviewed by a Variety Release Committee, including representatives from various institutions and groups, but in most cases dominated by the public sector.

Where private breeding is widespread, such data on variety performance may come from official (co-ordinated) variety trials. The breeder thus has to present his selections to an official organization which performs or co-ordinates the V.C.U. trials and/or the D.U.S. trials. Often, the breeder has to pay for inclusion of his materials in these tests.

The above system has particular consequences:

○ The evaluation of V.C.U. trials normally places great emphasis on yield potential throughout all trial sites. This means that the variety has to perform well in many different ecologies, i.e. be widely adapted. There may be selections that can perform very well in a particular area within a country (they have specific adaptation), but they may not be released because the average performance over all sites is not exceptionally high.

○ Where funding of the trial system is poor and trials do not get sufficient attention, statistically significant differences are hard to obtain. This means that very few varieties show a sufficiently clear yield advantage, and very few are released as a consequence.

○ When variety release committees consist of high-level scientists and administrators only, the likelihood is that the committee does not meet frequently, leading to few and delayed releases. Secondly, the users of the varieties, the seed producers and farmers, may not be listened to. This may lead to over-emphasis on figures and less attention given to qualitative aspects of the varieties.

○ Where both public and private breeders are active, the system may support the former and delay the release of privately bred (and often imported) varieties.

○ A requirement for variety description according to strict D.U.S. principles results in increased attention to the uniformity of varieties, even where uniformity does not serve an agronomic purpose.

74

In many countries only seed of released varieties can be multiplied or distributed. In other countries, such rules apply only to particular (notified) crops. For any non-notified crop, the spread of varieties and the production of seed is uncontrolled.

Variety release systems can be very effective where they produce a lot of valuable data which can assist farmers to choose among the available varieties. The system also allows for identification of varieties (through the D.U.S. trials), and can restrict the use of poor varieties, such as those susceptible to important diseases.

Where the release system is inefficient, however, or where it results in a bias in favour of particular types of varieties (e.g. wide adaptation), the release system may severely restrict the farmers' choice of potentially useful varieties.

Seed quality control
Seed quality is commonly controlled as a measure of 'consumer protection', i.e. poor quality seed is banned from the market to avoid farmers having poor crops and low yields. As such, seed quality control serves the national interest as well. Seed quality control usually involves regulation on the following points:

○ Regulated variety maintenance and a generation system for multiplication with clear denominations such as Breeder's, Foundation, Registered, Certified Seed.
○ Field inspection: checking on varietal identity, varietal purity, isolation, agronomic practices, and diseases.
○ Seed sampling and testing: germination, purity, moisture content, and a check on the presence of seed-transmitted diseases.
○ The labelling of sealed containers.
○ Marketing control: checking the validity of labels and quality of the seed in the market.
○ Post-control: grow-out tests of certified seed lots to identify the variety and the number of off-type plants.

These functions are performed by the (public) seed production organization, or in a further stage of the development of a formal seed system, by an independent seed certification and/or control agency, often a department within the Ministry of Agriculture. Especially in the former case it may be difficult for a private or voluntary organization to have access to seed quality-control services. Seed produced by farmer groups outside the certification system may not be tested in a formal seed laboratory and the seed-production experience of seed inspectors may not be available to them.

Seed laws are often quite specific with regard to seed quality control requirements. Many do not make a specific exemption for seed produced by farmers for local use. The definition of the word 'seed' determines the extent to which the law applies, i.e. which seeds have to undergo the centrally performed quality checks. Laws may refer to 'all plant materials meant for sowing or planting', or to 'all seed which is sold, bartered or otherwise changes hands', or to 'all seed which is sold in sealed and labelled containers'. The former two definitions include any seed, even the seed saved by farmers (the first definition), and the seed which a farmer gives to a neighbour or relative. It may be clear to the legislator that certification of all seed is impossible, and that farmers' varieties cannot be regulated. Farm-saved and locally bartered seed is therefore hardly ever restricted in practice. Strict regulation may, however, restrict the development of small-scale seed enterprises or co-operatives. The seed produced in such initiatives may have to comply with all procedures and standards of certified seed from the very beginning, which de-motivates groups to venture into producing better seed. The last definition of 'seed' leaves a lot of options open for local producers, but even then, the strict implementation of the law may hinder useful developments.

Organization
The organizational aspects of seed laws normally prescribe the formal institution of a national seed board, a certification agency and some technical committees, such as the Variety Release Committee.

4.2.3 Possible effects of seed laws on local seed initiatives
In practice, local seed production by farmers is hardly ever hampered by these regulations. In seed development initiatives aimed at supporting (unregulated) farmers' seed production practices, however, there is a chance that organizations will experience conflicts with the seed law. This may cause considerable problems. Local seed development initiatives may use local unreleased varieties and the formal distribution of such varieties may be prohibited unless they are formally registered. There may thus be a request to offer seed for formal trials. Such varieties may then not outperform the standard varieties under the testing conditions, or they may not be sufficiently uniform for formal release, with the result that the initiative may be blocked. In addition, if the seed certification agency makes seed testing, labelling and even a formal generation system obligatory for seed from the small-scale projects, this may turn out to be very costly and bureaucratic.

In many cases the government does not interfere with seed development initiatives, at least not in the initial phases. When such initiatives become successful, however, national seed corporations or private seed companies may object to these activities. They may experience competition in the market and mobilize the seed certification authority. Another reason for interference is political; i.e. political opponents of a particular NGO which is active in seed development can frustrate their activities through legal procedures.

When planning seed development initiatives, organizations should acquaint themselves with the national regulations. It may be wise to explore the attitude of the authorities towards activities such as the local promotion of unreleased varieties, assistance to local non-certified seed production, or the development of small-scale commercial seed supply. Early contact with the authorities may avoid serious frustrations. On the other hand, if political support cannot be generated, early contact may result in seed initiatives being frustrated from the very beginning.

4.2.4 Phytosanitary laws

Phytosanitary laws are meant to avoid the introduction of new pests and diseases into a country. They regulate the importation of seeds, food grain and any other materials which may contain organisms harmful to plants. Imported seed lots have to be presented to the phytosanitary authority upon arrival in the country. They can then be sampled and tested for the occurrence of listed prohibited pests or diseases. The seed may be kept in bond during such investigations. Sometimes, an internationally acknowledged phytosanitary certificate will suffice for the quick release of such seed lots. Rules for the importation of vegetative planting materials, such as potato, cassava, rubber or sugar cane, are generally very strict compared to those for true seeds.

Contrary to the seed law, which may be interpreted liberally for reasons given above, phytosanitary laws should be taken seriously. Researchers and development projects can cause considerable damage by importing seed samples in good faith. When local seed initiatives include the importation of seed for testing, the organization should make sure that it does not import diseases as well. When national procedures for quarantine are very time consuming and bureaucratic, it may be useful to ask for a phytosanitary certificate from a reputed laboratory in the country of origin to accompany the seed. In most countries this allows for an unhindered importation. In some cases this means that importation has to go through a laboratory in a third country. It is quite common that seed samples brought from Latin America into Africa are shipped through a European research institute for sampling and testing. Knowledge of such national regulations is important for any importer of seeds.

4.2.5 Plant Variety Protection: plant breeders' rights

Plant Variety Protection (PVP) aims at giving breeders the chance to earn money on the varieties that they developed by allowing them the exclusive rights to commercialize them. This means that a breeder is the only one to produce and sell seed of the variety and he or she may license others to do so, commonly in return for the payment of a royalty. PVP laws commonly include a so-called farmers' privilege and a breeder's exemption. The farmer's privilege implies that a farmer is allowed to produce seed of a protected variety in a non-commercial way without the consent of the breeder, i.e. for his own use and sometimes also for exchange or sale of small quantities. PVP was developed in Western Europe (see Box 4.1) and has spread to a number of other countries since. The breeder's exemption means that any protected variety can be used by a breeder to produce a new variety. The 'owner' of the protected variety does not have any rights to the new variety resulting from such use.

Box 4.1 The development of plant breeders' rights in The Netherlands

In the 1920s, wart (Synchytrium endobioticum (Schilb.)) was a serious problem in potato production in The Netherlands. At this time, the Institute of Plant Breeding of the Ministry of Agriculture in The Netherlands produced a list of varieties of agricultural crops in The Netherlands to inform farmers about available varieties and their performance. To establish the value of the varieties in the list, the Institute used information provided by private potato producers with a mandate in breeding, the Institute distributed seeds or seedlings from crosses between popular and wart-resistant cultivars to hobby-selectors. The Dutch seed certification service organized fairs and exhibitions at which successful selections were awarded. However, it was felt that these awards did not sufficiently reward the work of the farmers who had selected them; the seed industry and the farmers were already making large profits from these varieties. This concern has resulted in the creation of a fund from which the farmer-breeders (including the hobby-selectors) were payed a premium for each hectare planted with their selection. This initiative was the basis for the first law on Plant Breeders' Rights in the Netherlands, which became effective in 1941.

PVP may restrict local seed development initiatives. The 'owner' of a variety may not only charge royalties on seed produced, but he may even prohibit the multiplication of his varieties. It may not be feasible for the owner to monitor all seed produced (the cost may outweigh the benefit), but when a seed development initiative using a protected variety becomes popular, the owner is likely to claim his rights. This may result in the obligation to pay royalties, which is in principle justified, but it can also be possible that the breeder prefers to protect (his own) formal seed production by prohibiting the production of his variety by others. In the latest international convention on Plant Variety Protection, the rights even extend to the harvested product if the seed was obtained without the consent of the breeder. This allows the owner to execute his rights in the grain or vegetable market.

It must be stressed that a breeder deserves sufficient compensation for his work. This will give the breeder the financial capability to invest in further breeding, which should benefit farmers and consumers. It is therefore a good policy for seed production projects that use protected varieties to negotiate a deal with the owner. In practice, the amounts of money involved in getting the approval should be low, since without such a deal the owner would have to spend large sums on 'policing' the use of the variety.

4.2.6 Related issues

Plant Variety Protection (PVP) is a property rights system which has been specifically designed for plant varieties, because it was found unethical or impracticable to grant patents on living organisms. With the development of biotechnology, however, patents have entered the world of biology. Patents give a much stronger right to the holder; there is no farmers' privilege or breeder's exemption in the patent system. Patents may restrict the use of protected materials (or their genes) and may thus severely restrict the options available in participatory breeding programmes. Participatory Plant Breeding (PPB) tends to use a broader genetic base than modern plant breeding and since PPB often results in diverse varieties or the recommendation of different varieties for specific uses, a patent holder would be unable to monitor the use of his protected genes or plants, and may thus refuse the use of his materials in such programmes.

The World Trade Organization promotes the spread of Intellectual Property Systems worldwide. The Agreement on Trade Related Aspects of Intellectual Property Rights (TRIPs) of WTO prescribes the inclusion of patents and other IPR systems in national laws of WTO member states. For living organisms such as plants and animals, countries may design a separate, so-called *sui generis* system. *Sui generis* means 'of its own kind', i.e. in terms of the WTO an intellectual property right other than a patent (such as, for example, PVP).

Plant breeders' (PVP) rights and patents on materials or genes derived from farmers' developed and maintained landraces are hot discussion points in international debate on the rights of farmers and equity of access. The central question is whether farmers should pay for varieties of which the genes were selected from local varieties; farmers have developed and maintained these genes themselves (see Box 4.2). This consideration is given attention in the international debate, parallel to the formulation and implementation of PVP and patent regulations. There have been several efforts ethically and economically to balance the interests of industries and communities (also translated in terms of North–South, powerful and powerless, and so on).

One way is to value the breeding work of many generations of farmers who have domesticated crops and developed the 'building blocks' for modern plant breeding: landraces and other genetic resources. This is contained in the concept of farmers' rights which was formalized by a number of international agreements designed by FAO. Rules and infrastructure for this concept, however, still need to be designed.

Box 4.2 The rights of others on products derived from farmers' materials

In 1977, the International Institute of Tropical Agriculture (IITA) in Nigeria gifted an American scientist, A. Gatehouse, on a scientific mission in West Africa, with dried cowpeas. Several years later, Agricultural Genetics, a biotechnology company which the scientist had joined, patented an 'invention' named CpTI gene or Cowpea Trypsin Inhibitor. Through the modern techniques of biotechnology, the scientist and his colleagues had successfully extracted the disease-resistant genes from the African cowpea (later known as the variety TUV 2027) and injected it into different crops, such as soybeans and maize. Agricultural Genetics is projected to make hundreds of millions of dollars from this 'invention'.

Adapted from: RAFI Communiqué, 1990.

A further agreement is the Convention on Biological Diversity, which is instrumental in supporting the claims by nations to protect their genetic resources. It is generally agreed that nations have sovereign rights over their plant genetic resources, and the duty to protect these. However, this international agreement is only partially translated into action in most countries. Finally, genetic resources may be brought under the Traditional Resource Rights systems of the United Nations.

At the time of preparing this book (autumn 1999) the different claims on rights have not been brought under one broadly accepted system. Major problems include:

○ Who will grant a right?
○ Who will have a right (person, community, or nation)?
○ On what will rights be granted (population, stable variety, characteristics, genes)?
○ Who will determine the value of protected material?
○ Who will distribute the revenues?

Despite the general agreement that farmers have contributed and that genetic resources belong to a nation's natural resources, there remains a wide gap of political, technical and legal problems still to be solved.

4.3 Contributions by the formal farmer support services

The conventional wisdom concerning the function of agricultural services (credit, extension and input supply) with respect to the formal seed sector is that the extension service has to persuade farmers of the benefits of using modern seed; that agricultural credit must be available to enable farmers to purchase the seed; that farmers need a range of complementary inputs (particularly fertilizer) in order to make maximum use of modern seed; and that smoothly functioning distribution systems for these must be in place.

4.3.1 Extension

In this 'conventional wisdom', the extension service performs demonstrations of modern varieties and formal seed, and is often an important distribution channel for public and sometimes also commercially produced seed.

The alternative to this opinion on the role of the extension service is that, as long as using modern seed brings real, reliable benefits within the small-farm farming system, extension advice is irrelevant, because most farmers can obtain all the information they require from within the community. Various studies show that extension advice cannot provide the motivating force for adoption if the seed technology on offer is technically and/or economically irrelevant to the small-farm farming system.

The question, then, is whether the formal extension system has anything to offer in supporting local seed systems.

Even though extension service organizations in many countries display serious flaws, grassroot extension agents could be very effective supporters of seed system development. When extension agents are properly sensitized about the values and limitations of both locally and formally produced seed, they can be important collaborators. Variety demonstrations can be turned into participatory variety evaluations under farmer conditions and including both modern and local varieties. Extension agents can also be mediators between local seed groups and the official seed-quality control agencies, and extension agents can distribute good (local and modern) varieties beyond social and ethnic borders.

For this function to operate effectively, however, grassroot extension agents need support from the system that tends to concentrate on high-production agriculture only. Local seed system support activities have to take these limitations of extension agents into account.

Another very important function that the extension service can have is the channelling of information in the opposite direction. Information from the farms to the researchers and formal seed suppliers can significantly assist breeders to prioritize their breeding objectives, and advise seed producers about the attributes of seed that farmers really need. Theoretically, extension services are in a strong position to fulfil this function as they are in direct, independent contact at grass roots level with a large proportion of the potential client group for improved seed. In practice, however, budget restrictions and a lack of incentives can limit the effectiveness of the extension service in this respect. Local seed system support groups could help

81

extension agents by supplying data that support the case at the research and policy levels.

4.3.2 Credit
The view that credit is essential for small farmers to be able to use improved seed is entrenched in many government departments of agriculture. In practice, however, this view is debatable. The evidence from surveys in Pakistan and Kenya is that the availability of credit is not, in fact, an important variable explaining the adoption of formally produced seed by small farmers compared to others such as agro-climatic zone, farmer education, etc. (see Cromwell, 1990; Cromwell et al., 1992).

Two factors are significant here. First, modern seed accounts for a very small proportion of total production costs within most farms. Second, the available evidence suggests that, contrary to the usual assumption, modern seed such as the formal seed sector supplies can often provide a production benefit, albeit limited, even without the more expensive complementary inputs with which it is usually packaged. Where varieties do need such complementary inputs, careful economic analysis of the costs and benefits is extremely important. Where recurrent investments in seeds have to be made, such as in the case of hybrids, this is even more necessary.

Credit can be very important where farmers or farmer groups plan to get involved in specialized seed production. Seed drying, cleaning and packaging facilities, even locally adapted designs, require an investment that the groups need both financial and financial management support for. Local seed support activities may supply such resources or may mediate between the farmers and the banks. It may even be necessary initially to subsidize the price of seed. For example:

○ where project start-up costs have been very high;
○ where other agencies are continuing to provide free seed (for example, under seed relief programmes); or
○ where the local seed enterprise is producing seed of local varieties, but is facing competition from subsidized modern varieties distributed by government agencies or other donors.

A long-term problem may then arise when local seed enterprises which receive agency funding may not be able to sustain the relatively high production costs. Thus, if it is known at the outset that agency funding will eventually be withdrawn, then it may be more appropriate to strengthen local seed security using other approaches, rather than local seed enterprise development.

4.3.3 Inputs
Inputs such as fertilizers and pesticides may produce better crops and higher yields. Because seed crops should 'look better' than ordinary food crops, the

use of inputs in seed production is generally high. Seed system support activities should require inputs only when careful analysis has shown that the use of the input indeed solves an acute problem. For example, the spread of a seed-born disease might be curtailed through field sprays or the use of a seed-dressing chemical, or seedling vigour might be improved through the application of phosphate or potassium fertilizer in deficient soils.

4.4 Contributions to local seed systems at the policy level

4.4.1 Economic policies
Agricultural price policies influence seed production and distribution in various ways. Price policy on farm produce determines the value of the crop that is sold in the market, and thus the attractiveness of cash investments in each crop. Increased investments in seed could involve the regular purchasing of seeds, building better seed stores or giving seed crops special care. Conversely, where grain prices are kept artificially low to cater for urban consumers, for example, it may be impossible to invest in seed quality.

For subsistence crops, such investments are unlikely to be made. This is the case for pearl millet, which is a subsistence crop in Zambia without a seed market, whereas in India, where large quantities of millet grain are marketed, there is a ready market for hybrid seed of the same crop.

In many developing countries, the government frequently intervenes in price-setting of seeds to ensure that development objectives are met. Typical development objectives include:

○ encouraging farmers to use seed of modern varieties in the belief that this will increase productivity, and improve household and national food security;
○ fulfilling macro-economic policy objectives with regard to agricultural production and cropping diversity;
○ ensuring adequate income for participants in the sector, especially contract growers, small-scale processors and small traders; and
○ ensuring sufficient supplies of seed.

The exact balance between the conflicting objectives of cost-recovery and equitable seed distribution is often poorly defined and results in poor performance on both counts. Public subsidies of seed can, however, seriously discourage the development of more sustainable local commercial seed production.

In addition to seed-related regulations such as those presented in Section 4.2, general macroeconomic laws, such as those on investment, tax and labour, affect the seed sector, especially the emergence of small seed enterprises. Investment and exchange regulations are likely to influence competition in the formal seed sector and thus its effectiveness to serve farmers. Where value-added tax is charged on seeds, informal seed supply is promoted. A government thus has a wide range of tools to stimulate or restrict different developments.

4.4.2 Research policy

Agricultural research is commonly regarded as a national task, which supports the change from subsistence farming to market-oriented agriculture. In colonial days, research was primarily geared to supporting the industrial and export food crop sectors, but started to concentrate on food crops in order to feed the urban population. Where this is still the leading principle, little can be expected from the agricultural research system with regard to breeding for marginal conditions, breeding subsistence crops, and participatory plant breeding in general. The development of other seed-related technologies is likely to happen only under research policies that specifically address the needs of low external-input farming.

During the 1980s there was a growing awareness that not all agricultural land could be put under high-input 'economic' farming. In the current trend towards privatization, however, there may be a move towards 'doing research for those who can afford to pay for it', i.e. in most cases the so-called 'economic' farmers and the food processing industries. Research policies are thus ambivalent about their orientation towards farmer participation and a diverse agriculture.

4.4.3 Seed policy

Different policies affect the development of the seed sector. It is therefore very important that a country defines a comprehensive seed policy, in which the tasks of the different players in the sector are well defined. Such a policy would take a balanced approach to formal and informal seed supply, giving guidance and support to those with commercial and developmental intentions for developing the availability and quality of this basic input for all agriculture.

A seed policy should take into account the differences between 'seed products', whereby for some crops seed production is likely to be commercialized in large companies (the main vegetable crops, hybrid cereals, etc.); for some crops smaller-scale commercial enterprises may be expected to emerge (self-fertilized cereals and pulses), and for some crops and farming systems the quality of farm-saved seed is the primary concern, because commercialization cannot be expected to take place at all (subsistence crops). Such a policy should also balance different concerns, such as the modernization of agriculture and national food security, a free market orientation, care for the rural and urban poor, regard for environmental and biodiversity protection, international agreements, and so on. It is important that farmers are well represented in committees that prepare such seed policy and seed strategy documents. This representation should include both economic and subsistence farmers.

PART B

TECHNICAL ASPECTS OF
SEED PRODUCTION

5. Seed quality

Seed is a necessary input for all agriculture. In discussing technical aspects of seed system support, quality is the guiding principle. Any support to seed systems necessarily deals with the four quality aspects which are introduced in Section 5.2: physiological, analytical, sanitary and genetic seed quality. The need to concentrate on quality is irrespective of the strategy of seed system support chosen in any particular situation: whether improving farm saving of seed, assisting community efforts in agricultural development, supporting the development of small seed enterprises or improving formal or semi-formal seed production. With this focus on seed quality the methods to improve seed supply through technical means are introduced in the remainder of Part B.

5.1 The significance of seed quality

Seed is a necessary input for all agriculture. There are two major aspects to 'seed' in a farmer's view: *quality* and *availability*. Strategies to assist seed systems have to take both these aspects into account. Aspects of availability have been dealt with in the sections covering sources of seed and seed security (Sections 2.4 and 2.5). In discussing technical aspects of seed system support, the quality aspect is the guiding principle. Any support to seed systems irrespective of the strategy, necessarily deals with the four quality aspects that are elaborated below.

In relation to quality, farmers' primary concern is to grow vigorous and healthy crops. Apart from many other agronomic concerns, this means that he or she needs seeds that germinate and produce vigorous seedlings. Secondly, the seed determines the spread of certain diseases, and the important varietal aspects of the crops such as adaptation to local cropping patterns and environmental conditions, yield potential, and product quality. Throughout the following chapters these quality aspects will appear in discussing seed production, post-harvest operations, selection and market options. These sections take local seed production techniques as a starting point, but 'borrow' from knowledge and technologies that have been developed in more formal seed systems.

5.2 Aspects of quality

Four basic seed quality aspects can be distinguished:

○ physiological quality (germination, vigour)
○ sanitary quality (absence of seed-borne diseases)
○ analytical quality (percentage of good seed in a particular lot)
○ genetic quality (varietal adaptation, varietal purity).

This section discusses these four aspects in relation to local seed system development. Obviously these quality factors are interrelated; however, distinguishing between these factors helps to analyse the seed-related limiting factors in agriculture, and to plan actions accordingly.

5.2.1 Physiological seed quality

The first basic requirement of seed is that it has to *germinate* at the right time. Secondly, the seedling has to be strong enough to withstand the environmental conditions that it faces when emerging. The young root system has to be able to support the plant, the young stem should be able to break the soil and the first leaves should not wither during the first sunny day. The ability of the seed to produce a strong seedling under farming conditions is called *vigour*.

Seed of low physiological quality will result in a poor crop development and a low plant density. A low germination rate is not necessarily a great problem, however: when a farmer knows before sowing that the germination rate is, for example, 50 per cent, the seed rate can be doubled to arrive at a good plant density and a good yield. The problem is, however, that a seed lot with such low germination capacity normally also exhibits low seedling vigour, which means that the development of the crop is slow, giving weeds more chance to develop, and that many plants are too weak to withstand stresses such as an early drought period, or disease attack. Also, the use of larger quantities of poor-quality seed results in irregular crops. A low plant density does not necessarily result in low yield. In some crops, such as sorghum, tillering of the remaining plants will to a large extent make up for a lower number of plants, when the open spaces have not been filled by weeds. This ability to compensate is lower in other crops, such as maize or groundnut. An 'open' stand can reduce yields of groundnut greatly because aphids that spread the Rosette virus prefer open crops to a well-established closed canopy, thus causing more damage in the former.

Seed size

Seed size is not the best parameter to establish physiological seed quality. The largest seeds of legume crops like beans and soybean are often 'soft', i.e. they do not give the strongest seedlings that can resist stresses well. On the other hand, smaller seeds can be very vigorous, but sometimes lack the strength to reach the soil when planted too deep. Shrivelled seeds are of course never the best bet. In general, the density of the seed is more important than size, i.e. plump, well-filled seed of average size are usually of good quality and the best option.

Storage
Most fully developed seeds on a healthy plant are in principle able to germinate. From the moment that seeds start to dry on the plant they start an ageing process: the capacity of the seed to germinate gradually decreases. Moisture and temperature are the main factors in this deterioration process. This means that seeds should be harvested early, dried well, and stored under dry conditions. Secondly, seeds should be protected from over-heating during drying and should be stored as cool as possible. Proper drying of the seed after harvesting does not solve the problem for good; seed as a living organism can also absorb moisture during storage. When the humidity of the air is high, the seed moisture content will rise (see Chapter 7).

Dormancy
There are conditions in which seeds are alive and healthy, but do not germinate. This aspect is called dormancy. Dormancy is a system that prevents plants in natural populations from germinating simultaneously, and is useful to bridge an unfavourable period. Dormancy 'spreads the risk' of germination and is thus important for the survival of a population of plants. In agriculture, however, the farmer wants all seeds to germinate soon after planting. Dormancy can thus be a serious problem in agriculture.

During the domestication of crops, farmers' selection has changed this genetically controlled dormancy in many crops. Total absence of dormancy may not, however, be very beneficial either; germination of the seeds before harvesting (i.e. on the ear, in the pod or cob) or early sprouting in storage (potato seed tubers) should be avoided. This early germination on the plant before harvesting can be observed in different crops (e.g. barley, groundnut). The importance of dormancy is illustrated with the groundnut crop in Box 5.1.

Dormancy mechanisms include mechanical barriers: a seed coat that is impermeable to water (bean) or oxygen (beet), or just too hard for the embryo to break (rose). Seed dormancy can be also be regulated by the presence of hormonal inhibitors in the seed coat (*Fraxinus*) or because the embryo finalizes its development only after soaking (palm trees).

Breaking dormancy is normally not a problem for traditional crops in any area. The dormancy level of e.g. groundnut varieties is commonly adapted to the prevailing farming system. Dormancy can cause problems in introduced crops and varieties, however, and in semi-wild crops like pasture legumes. Breaking dormancy can be rather difficult. Sometimes, simple soaking in water may wash out the dormancy-inducing compound from the seed coat, or rubbing the seeds with sandpaper can reduce the mechanical barrier or allow the seed to absorb moisture. More often, however, hot water (beans) or sulphuric acid (*Hibiscus*, lettuce) are necessary.

Box 5.1 Dormancy

Dormancy is very important in groundnut, where seeds grow in an environment soil which is favourable for germination as soon as moisture conditions are optimal. Without dormancy groundnuts could germinate very easily on the plant when harvesting is delayed, even by a short time.

Important groundnut varieties belonging to the Virginia group have a marked dormancy. These are produced in areas with only one season per year and thus a relatively long storage period of the seed, in which dormancy is gradually broken. When the seed store is dry, the seeds will not germinate until planted. For those groundnut-producing areas with two cropping seasons, dormancy can be a serious problem. When the seeds do not germinate for the second cropping season, i.e. often within a few weeks or even days of the first harvest, a good second crop is impossible. In such areas, groundnut varieties of the Spanish-Valencia group can be planted. These varieties do not have a marked seed dormancy. Harvesting at the right time is, however, essential for these crops.

5.2.2 Sanitary seed quality

The absence of seed-borne diseases is another important aspect of seed quality. Most diseases that limit crop production spread through spores that are carried from plant to plant by wind, water or animals. Some remain in the soil or plant debris, others survive in weedy host plants during the period of the year when there are no crops in the field. A number of diseases can, however, be carried in or on the seed. Such diseases may negatively affect germination or vigour and damage the following crop. Absence of such seed-transmitted diseases is therefore an important seed quality characteristic. These seed-related diseases include fungi (e.g. smut in sorghum), bacteria (common bacterial blight in French beans) and viruses (mosaic virus in cucumber).

Such diseases may be present in different parts of the seed, such as the endosperm, the embryo or the seed coat. Other diseases may attach themselves to the outside of the seed, or will form structures that may mix with the seed. The options for getting rid of them easily depend on the seed structures in which the pathogens are present. (see Section 6.3).

5.2.3 Analytical seed quality

For clients of commercial seed producers, the quantity of good seed in the bag is a very important aspect. Mechanical harvesting and threshing often result in a mixture of fair amounts of straw or other plant parts and lumps of soil with the seed. The presence of lumps of soil or straws may not be considered a major problem by farmers who produce seed for their own hand planting; such 'inert matter' only blocks planting machines. It is, however, important for farmers who purchase seed, and who do not want to pay for a large percentage of useless material.

Another important aspect of analytical seed quality is the possibility of the presence of weed seeds in the seed lot. This is particularly important where seeds are carried over some distance to areas where the particular weed is absent or of minor importance, but even in already infested fields the increase of a weed problem should be avoided. Sometimes the seeds of such weeds have the same size and shape as the crop seeds, which makes them difficult to remove (e.g. red rice in paddy, *Cuscuta* in Alfalfa, wild oats in wheat or barley), or the seeds of other weeds are so small that they can hardly be observed, or they stick to the crop seeds (e.g. witchweed in cereals) (see Box 5.2). Special machines have been designed to clean these mixtures of seed, weeds, lumps and straw (see Chapter 7).

Box 5.2 Weeds: the red rice case

The expression 'weed' is not always clear cut. Red rice is a good illustration of this.

Surinam
Mixtures of red- and white-coloured rice considerably reduce the value of an export rice crop. Controlling red rice in the paddy fields of Surinam is one of the most important aspects of agronomic research. Mixtures of red and white seed can be avoided in seed production through the combination of a proper generation system in seed production and laboratory analysis. Red rice is not only a problem because the seeds are harvested with the crop seed, and are difficult to clean with standard seed cleaning equipment; it also drops a portion of its seed before the rice crop is harvested. Red rice thus becomes a very serious weed, even though the seed is edible just like other types of rice.

Philippines
In many places in South-east Asia red rice is desirable: it is considered nutritious and fetches a good price in the local market. Occasionally, red rice plants appearing in paddy fields are therefore selected by farmers and the seeds are planted separately to see whether they will develop into a new local variety of red rice. In this case the 'weed' has an important value.

5.2.4 Genetic seed quality

The genetic quality of the seed determines to a large extent the success of the crop. The genetics of the seed determines important aspects such as yield potential (i.e. the yield under optimal growing conditions), the tolerance to stresses such as drought, waterlogging, frost, low fertility, and resistance to diseases and pests. It also determines characteristics related to food-processing (cooking time, milling), local tastes or market requirements.

Varieties
Varieties may differ very much in their adaptation to particular growing conditions. These include environmental conditions, such as soils, rainfall, and length of the season, but also farm management aspects, such as soil preparation, weeding and harvesting method. The choice of the right variety is a major concern for every farmer.

Uniformity
Where modern varieties are appropriate, the farmer likes to have few 'off-type' plants, i.e. plants that clearly do not belong to the original variety. This is what is meant by varietal uniformity. Such varietal uniformity is determined by the genetic quality of the seed. Varietal uniformity is usually important when crops are produced for the market, i.e. where uniform products may fetch a better price. An example is common bean in many parts of Africa, where the market generally prefers uniform-coloured produce to mixtures. Uniformity is also important for vegetables in many markets. For export markets, uniformity is often an extremely important requirement. From an agronomic point of view, other aspects of uniformity are important: uniformity in plant height reduces uneven competition among plants; uniformity in maturity period allows for a single harvest. These aspects are particularly important for farmers who can control the major limitations in agriculture, such as water and soil fertility. Farmers in more marginal conditions, however, may benefit from uneven ripening of the crop and from a mixture of plants in the crop.

Changing seed quality with time
The genetic composition of varieties and landraces may change with time (see also Section 1.1 and Box 2.2). This can happen as a result of mutation, introgression and the mechanical mixing of seeds. Cross-fertilization occurs in virtually all crops, but in some crops much more commonly than in others. In so-called self-pollinating crops, it is rare. The variety then breeds 'true-to-type'. Examples are most legumes and cereals like wheat and finger millet. For these crops it is easy to keep the original variety intact. For cross-fertilizing species, however, special precautions may have to be taken to avoid major changes in the variety which might affect their adaptation. Examples are maize, pearl millet, sunflower and cucumber.

Natural changes in varieties are commonly referred to as 'degeneration of the seed', or 'the seed gets tired'. Such degeneration is the result of a lack of selection by the farmer. Seed also gets 'tired' during a gradual build-up of seed-transmitted diseases. Seed size and crop yields gradually decrease and often the value of produce declines due to discoloration. Another reason for seeds to degenerate can be explained by selection pressure. A variety is selected by farmers for characteristics that she or he likes, such as large seeds. Natural selection may work in the opposite direction: 100g of small seeds produce more offspring (are more fit) than 100g of large seeds. When the farmer does not apply selection for large seeds over a number of seasons, the variety may degenerate. Obviously such natural selection occurs more in heterogeneous varieties of cross-fertilized crops than in uniform self-fertilized crops. Another reason for degradation may be the gradual build-up of mutations in the variety, which can introduce negative characteristics into the crop (see also Section 1.1).

Formal seed suppliers and extension agents commonly say that farmers should purchase certified seed every three or four seasons (the ideal seed replacement rate). This is far too general advice, because the degeneration depends on the presence of seed-transmitted diseases, the mating system of the crop and the characteristic of the variety in relation to natural selection pressures. Above all, however, the rate of degeneration depends on the farmers, who may avoid diseases, and re-select the variety every season.

5.3 Monitoring seed quality

Monitoring seed quality essentially means that a regular and critical check has to be carried out of the seed fields, of the harvesting equipment and procedures, of the cleaning and storage operations, and of the conditions at retail outlets and shops. This monitoring serves two purposes: firstly all people involved are stimulated to give the seed extra care, and secondly reliable data can be collected for appropriate and timely management decisions. Such measures include timely roguing of weeds and diseased plants, protection of the crop by spraying, optimization of drying procedures, construction of adequate storage facilities, regular turning of stacks, back-drying, fumigation, rodent control, etc. Although seemingly obvious, this extra care for seeds is often lacking. Careful monitoring of the insect population and diseases is the basis for integrated pest management: it may enable farmers to postpone sprayings or even to skip sprays altogether. Careful monitoring of seed quality is the basis of good seedmanship. Section 10 presents the basics of seed testing and Section 12.5 some hints for experimentation.

6. Seed production agronomy

Important improvements in seed quality can be achieved by supporting appropriate agronomic procedures, starting with land preparation and the choice of seed, and ending with the handling of harvested materials. This section builds upon the concepts of seed quality introduced in Chapter 5 by discussing measures in seed production that may differ from normal cultivation practices. Such measures are presented as 'good practice in seed production', i.e. in the production of seeds as a specialized undertaking. The same issues are, however, important in combined (crop and seed) production systems, and can assist farmers and technicians to improve on their seeds.

6.1 General principles

6.1.1 The similarities and differences between crop and seed production

In general, the conditions and cultivation practices that lead to good yields also lead to good seed and a good seed yield. Each crop and ecological condition are different and require particular management decisions for seed production and handling. This book thus cannot give 'recipes' for cultivation and seed handling. The general recommendations in this section are meant to increase understanding of the general principles of seed production and will allow farmers, extensionists and development agency staff to improve their seed production. This section should thus be read in conjunction with literature on agronomy destined for the particular region, and be combined with local knowledge on crop cultivation and local seed production. Seed production is presented as an activity that is separate from crop cultivation; however, when only a small portion of a certain plot is used for seed, the suggestions made here are no less applicable.

A major principle of seed production is that in order to produce good seed a very good 'mother' crop is needed. Since seeds are commonly formed during the final stage of plant development, the crop has to be well nurtured until the last day. If during this final stage the mother plant is in sub-optimal condition, seeds may remain small and poorly filled.

A poorly managed seed crop may give very considerable yield or quality reductions in following years. This is the basic reason for putting an extra effort into caring for the plants that will produce seeds. For example: plants suffering from diseases may not only produce lower yields, but also develop smaller, less vigorous seeds, and this can jeopardize the establishment of the plants growing from these seeds in the following year. If the disease is carried over with the seed, the following year's crop may be seriously contaminated, resulting in large reductions in productivity; diseases may even contaminate the following crop and cause massive yield and quality reductions.

Broadly, two groups of crops can be distinguished when discussing seed production. There are crops which are grown for what is actually the propagation material, i.e. the seed, such as cereals, legumes, most oil crops and Irish potato. For these crops, farming practices for seed production generally follow the standard production methods for the crop. These include proper land preparation, optimum plant spacing and soil fertility management, and good weed, disease and pest control. The main differences between grain production and seed production concern the quality requirements of the latter.

By contrast, there are some crops that are not normally grown for their propagation material, and for which seed production can be a very specific activity. This group includes many vegetables, pasture crops, and many vegetatively propagated crops and fruits. Seed production in such plants may be easy, i.e. when planting materials are always available as a side product of crop production (e.g. the vines of sweet potato and the stems of cassava), or may require very specific knowledge and technology which warrant the specialization of seed producers. This is the case in grafted fruit trees and biennial vegetable crops, i.e. crops that produce a vegetative (consumable) plant in the first year and bear flowers and seeds in the second year only (e.g. onion, cabbage).

Some crops may fall into both categories: Irish potato can be multiplied using tubers (the consumed product) or seed (True Potato Seed, or TPS); fruit trees by seeds or vegetative materials; onions by true seeds or dry sets.

Finally, the production of hybrid seed of any crop requires such special planting designs and crossing measures, that production is very different from normal production and difficult to carry out without the support of a specialized organization (often a seed company).

The following section analyses the general differences between crop and seed production from planting to harvesting with special reference to local (i.e. non-formal) seed production. Specific information about crops will be dealt with later.

6.2 Land preparation and crop establishment

The main elements of land preparation and crop establishment are the selection of the field (rotation, isolation), tillage, the choice of the seed, the timing of planting and the plant density.

6.2.1 The choice of field and isolation

The plot for seed production has to be chosen carefully. The land should be able to produce a good crop, because weak crops will produce poor-quality seeds. Fertile well-drained soils with optimum water supply should be chosen.

The rotation of crops is more important for seed than for normal crop production. Firstly, rotation reduces the risk of building-up seed-transmitted diseases in the soil. Secondly, without a rotation there is more risk of mixing varieties when volunteer plants emerge in the seed production field (i.e. growing from seeds that dropped from the plants during the previous seasons). This is particularly important when growing modern (uniform) varieties. Generally growing the same crop in consecutive seasons has to be avoided, and where this is not possible, for example with paddy rice, it is best to grow the same variety for seed as was planted the previous season.

Isolation, or distance to plants that may cross-pollinate the seed crop, is important for cross-fertilizing crops. This means that, depending on the level and the mode (by wind or insects) of outcrossing, a minimum distance should be preserved to the next field planted with the same crop (see Figure 6.1). Within this distance, weeds that may cross with the seed crop should be removed as well. Such cross fertilization will reduce the varietal uniformity and change the original variety, which is not usually the intention of seed production.

If planting different varieties close to each other cannot be avoided, for example when a farmer would like to maintain a local population while the

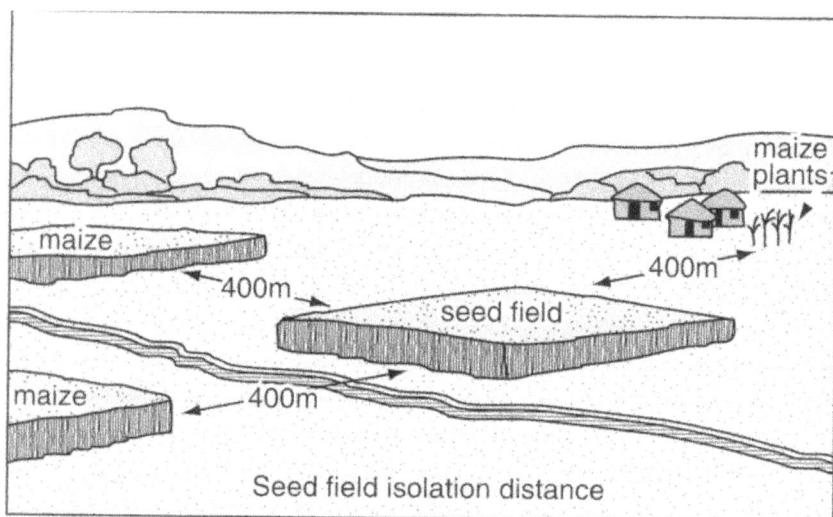

Figure 6.1 Isolation of a seed production field to prevent fertilization with pollen from a neighbouring field. For maize, the recommended isolation distance is 200–400m.

neighbour plants a modern variety, it is important that the seed field is large enough and that the centre of the field is harvested for seed, and not the outer rows facing the modern variety.

In some cases it is possible to avoid cross-fertilization through careful planning of the planting time. The simultaneous flowering of the seed crop and 'contaminating' plants can in some cases be avoided through such staggered planting.

However, the isolation requirement thus depends on the intentions of the farmer. When a relatively uniform variety is multiplied, isolation is very important, but when a more diverse variety is produced, some cross-fertilization from other sources may even be interesting to enrich the local variety and to allow for a continuous selection of better adapted types. Table 6.1 presents isolation distances for relatively uniform varieties of a number of crops. These distances should be regarded as a guideline rather than a definite requirement.

6.2.2 Environmental conditions
For some crops, seed production requires particular conditions, whereas the crops can be produced under a much wider range of conditions. Biennials like onion and cabbage are a clear example (see Section 6.4). Some other crops favour particular conditions as well: Irish potato seed tubers are best produced under cool conditions (over 2000m altitude in tropical areas). The cool temperature produces better-quality seed tubers, and even more importantly, in these conditions the presence of aphids that spread important virus diseases is reduced. Furthermore, some fungal diseases are less prominent at higher altitudes (lower temperatures). For crops that have such particular requirements, specialized seed production has also often developed within the local seed systems.

6.2.3 Tillage
For seed production, extra care may be needed to reduce the presence of weeds through proper land preparation: weeds compete with the crop for nutrients, weeds may be a host for crop diseases and weed seeds may mix with the seed during harvesting. Early land preparation may stimulate weed growth. When this is followed by harrowing just before planting, a lot of weeds may have germinated and been killed. Final preparation of the seed bed requires extra care to result in an even germination of the seeds, thus giving a uniform crop in which off-types and diseases can be easily spotted.

6.2.4 The choice of seed
The seed should have been selected from healthy plants of the required variety, preferably from an area where conditions are optimal for seed production (e.g. high altitude sources for Irish potato seed). The selection of seeds from a seed lot will be dealt with in more detail later. In formal

Table 6.1: Recommended isolation distances for seed production of relatively uniform varieties (based on formal seed systems)

Crop	Weight of approx. 1000 seeds[1]	Multiplication factor[1]	Isolation (m)	Fertilization mechanism
CEREALS				
Barley (*Hordeum vulgare*)	50	12	30	self
Finger millet (*Eleusine coracana*)	3	50	5	self
Maize (*Zea mais*)	300	80	300	cross (wind)
Pearl millet (*Pennisetum typhoides*)	7	120	300	cross (wind)
Rice (*Oryza sativa*)	25	20/80[2]	5	self
Sorghum (*Sorghum bicolor*)	20	100	100	semi-cross (wind)
Wheat (*Triticum aestivum*)	40	12	5	self
PULSES				
Bambara groundnut (*Voandzeia subterranea*)	500–750	12	100	semi-cross (insects)
Broad bean (*Vicia faba*)	500	15	100	semi-cross (insects)
Chick pea (*Cicer arietinum*)	500	20	5	self
Cowpea (*Vigna unguiculata*)	200	50	100	semi-cross (insects)
Lentil (*Lens esculenta*)	300	70	5	self
Pigeon pea (*Cajanus cajan*)	80	80	100	semi-cross (insects)
Groundnut (*Arachis hypogaea*)	300	8	5	self
OIL CROPS				
Sesame (*Sesamum indicum*)	3	120	300	semi-cross (insects)
Soybean (*Glycine max*)	150	20	5	self
Sunflower (*Helianthus annuus*)	40	80	600	cross (insects)

Crop	Weight of approx. 1000 seeds[1]	Multipli-cation factor[1]	Isolation (m)	Fertilization mechanism
VEGETABLES				
Beans (*Phaseolus vulgaris*)	400	10/20	5	self
Cabbage (*Brassica oleracea*)	3	1000	600	cross (insects)
Carrot (*Daucus carota*)	2	20	600	cross (insects)
Cucumber (*Cucumis sativus*)	40	100	600	cross (insects)
Eggplant (aubergine) (*Solanum melongena*)	3	1000	100	semi-cross (insects)
Bitter gourd (*Momordica charantia*)	250	100	600	cross (insects)
Melons (*Cucumis melo*)	150	120	600	semi-cross (insects)
Okra (*Abelmoschus esculentus*)	80	120	100	semi-cross (insects)
Onion (*Allium cepa*)	4	80	600	cross (insects)
Peppers (*Capsicum annuum*)	5	150	100	semi-cross (insects)
Pumpkin (*Cucurbita* spp.)	150	200	600	cross (insects)
Radish (*Raphanus sativus*)	12	200	600	cross (insects)
Tomato (*Lycopersicon esculentum*)	3	800	5	self

Note: (1) the figures for 1000-grain weight (grams) and the multiplication factor are general estimations only. Different varieties of beans, broad beans and many others differ considerably in seed size (and thus 1000 grain weight), and multiplication factors largely depend on yields which in turn depend on the potential of the area and the cultivation practices. Small-seeded varieties generally have a higher multiplication factor than large-seeded types of the same crop.
(2) The multiplication factor depends on whether rice is sown directly or transplanted.

seed production the general rule is that the quality of the seed produced can never be higher than the quality of the seed planted. This is the reason why such systems have a very strict *generation system*, with seed passing through different seed classes from Breeder's Seed to Certified Seed, in which seed quality standards gradually reduce.

In local seed systems there is no guaranteed source of high-quality seed; the farmer has to keep seed qualities high through good care and selection, or by obtaining seed from a farmer who is well known for his or her good-quality seed. For some crops, it is reported that changing the growing conditions may improve seed quality. This can be related to disease infection or physiological factors (as in the case of Irish potato).

6.2.5 The timing of planting

Early planting, i.e. at the onset of the rains, is normally recommended to produce strong crops and reduce pest and disease incidence. On the other hand, planting has to be planned in such a way that the harvesting time coincides with a dry season. In some cases this means a delay in the commonly used planting time. Harvesting in the rain without good seed drying facilities will cause quality problems. Ideally, a moderately fast maturation should be aimed at: too high temperatures during seed ripening may be damaging as well. Where late planting is necessary, but results in too poor crops, it may be worth planting for seed production off-season under irrigation. This will usually be more expensive, but it may be compensated through a better seed quality.

6.2.6 Plant population

Finally, the plant population is important. The plant population (number of plants per unit area) is a function of the number of the plants (i.e. number of seeds planted multiplied by the emergence rate) and the area planted. In general, row planting is recommended to allow for easier inspection of the seed crop and roguing (see later). Row planting also facilitates weeding and creates a distance between rows which can reduce the spread of some diseases.

For some crops the spacing for seed production is wider than for crop production (lower plant population). There are two reasons for this: firstly, the wider spacing between plants, and particularly between rows, can decrease disease incidence. Sun and wind can get into the crop, and the drier micro-climate reduces the incidence of fungal diseases; also, the sheer distance between plants can reduce the spread of such diseases. Secondly, the wider spacing allows each plant to develop fully. This can result in stronger plants, with higher seed yields per plant and plump seeds.

On the other hand, too low plant populations reduce yield. Furthermore, for some crops plant populations should be increased when the crop is grown for seed: crops that tiller a lot may develop problems at harvesting when planted too widely. When the main branch (ear, head) is ready for harvesting, the seeds on the side shoots may not be ripe yet. It may then be difficult to decide when to harvest. This can be a problem with millets, but also with some vegetables like carrot. With some other vegetables there is another reason for closer spacing. When, for example, aubergines, sweet pepper, cucumbers or melons are grown for vegetable consumption, fruits are harvested when the seeds are still immature. After harvesting the first fruits, the plant continues to grow to produce a new flush of marketable fruits. When grown for seed, however, the first fruits remain on the plant, thus reducing vegetative growth. These plants therefore require less space and can be planted closer together for seed production. Another option is to combine crop and seed production by picking the first fruits at the vegetable stage, and leaving the last ones for ripe seed. This may be economically advantageous but it may result in poor seed quality for two reasons: firstly, at the end of the season the plants may be weak and produce less well-developed seeds; secondly, by the end of the season the plant has accumulated all the available diseases, some of which may be seed borne. It is therefore strongly advisable to allow early developing fruits to mature to produce better seeds.

Finally, for root crops like sweet potato and cassava, the spacing can be decreased when the crop is grown for propagation material. A very closely spaced cassava crop produces many stems for planting and hardly any roots. This is done only when large quantities of planting material is required, for example in a campaign to replace virus-infested cassava with 'clean' planting materials in a large area (a valley or a district).

6.3 Crop cultivation and protection

The main aspects of crop management are mineral fertilization, pest and disease management and inspection on varietal characteristics.

6.3.1 Fertilizers

The land should be fertile to produce a strong crop. The fertility requirements for seed may, however, be rather different from those of crop production. The main differences relate to nitrogen and potash.

Nitrogen applications in the form of chemical fertilizer or manure should be lower compared to normal crop production and can best be given in a split application, i.e. at planting and just before flowering. The high availability of nitrogen produces vegetative growth and generally high yields, but may result in 'soft' seeds, i.e. large seeds that cannot withstand adverse conditions during storage or emergence (i.e. less vigourous). Strong vegetative growth may also delay the maturity of the seed, thus increasing risks of yield losses. These are the main reasons not to give excess nitrogen as a base dressing.

Sufficient potash (K) and phosphorus (P) on the other hand are very good for the development of hardy seeds. Phosphorus generally increases seed yield, and potash is particularly important for flowering. When deficiencies in mineral supply are observed it can normally not be rectified during the growing season, and fertilization has to be done at planting. Potash and phosphorus are particularly important in oil crop seed production. The proper identification of nutritional disorders is very important. Box 6.1 presents some general guidelines for cereal crops.

Box 6.1 Key to nutritional disorders in sorghum (most of these aspects are also observed in other cereal crops)

A1. Symptoms appear first on fully expanded leaves or are more severe on older leaves — see B

A2. Symptoms appear first on still-expanding leaves or are more severe on younger leaves — see C

B1. Symptoms begin near the leaf tip and advance towards the base; yellowing between veins rare — see D

B2. Symptoms begin as yellowing between the veins, sometimes together with the sides of the leaves — see E

C1. Symptoms begin as prominent yellowing or browning streaks — see F

C2. Symptoms begin as dispersed spots or broad yellowing bands or as generalized yellowing or dying of tip and margins; veins never prominent — see G

D1. Plant pale green; pale yellowing browning advance down the main vein — Nitrogen deficiency

D2. Plant dark green; dark yellowing advances down the margins; purple colours common — Phosphorus deficiency

D3. Plant dull grey-green; drought appearance; grey leaf tip and dying of the margins — Sodium chloride toxicity

E1. Plant dark green; pale yellowing between the veins near margins; browning of margins — Potassium deficiency

E2. Pale green-yellow; dark yellowing and rust-brown colouration between veins — Magnesium deficiency

F1. Plant pale yellow and streaked throughout the leaf; no browning — Iron deficiency

F2. Plant dark green; white and later brown spots between veins in lower parts of the leaf; leaves die — Manganese deficiency

F3. Plant dark green; pale yellow streaks in mid-section of the leaf; leaf tears easily — Boron deficiency

G1. General pale yellowing of the leaf; browning rare — Sulphur deficiency

G2. Plant pale green, appearing wilted; tips and margins yellowing and browning — Copper deficiency

G3. Plant dark green; torn and brownish spots near torn areas; malformed leaves — Calcium deficiency

G4. Plant pale green; white to pale yellow bands between margin and mid-vein at lower part of the leaf, turning grey and greyish brown — Zinc deficiency

6.3.2 Pest, disease and weed management

Pest and disease management include the control of weeds, animal pests and diseases.

Weed control is important in crop cultivation, but even more so in seed production. When weeds get the chance to develop, they will compete with the crop for nutrients, light and water, and both the crop and the seeds will be weak. Secondly, weed seeds may be harvested together with the crop seeds, thus potentially giving rise to excessive weed problems in the following season. Furthermore, insects and crop diseases may hide in weeds in or bordering the field.

A final reason for weed control is the possibility of cross-fertilization between crops and related weeds. An example is so-called 'wild sorghum' in sorghum fields. The plants look similar in the early stages, but the seeds are very small. When the weed and the crop cross, grain production is not affected, but the following generation will include plants that do not give a palatable harvest. Such crosses may be useful to add new characteristics to the cultivated crop, and such crosses are nurtured by particular individuals and communities. The usefulness of such crosses will, however, generally show only after a number of seasons. Other examples can be found in the cabbage/mustard family, which includes a number of weeds that may cross with crops.

In general, all good crop cultivation practices in relation to pests and diseases are important for seed production as well.

Pests and diseases generally reduce crop yields. The earlier a disease attacks a crop, the more it can affect it. Introducing a disease with the seed can therefore be very detrimental. The control of seed-borne diseases should therefore receive extra attention when producing a seed crop as compared to a normal crop. This requires a good knowledge of the symptoms and the remedies. These are crop and region specific and so cannot be dealt with in detail. Table 6.2 presents some major seed-transmitted diseases of some crops.

Leaf-eating pests can limit crop growth and reduce seed size. There are also insects that feed on the seed, such as the green stink bugs that suck on pods of soybean. Heavy damage results in total loss of yield, but limited damage to the seed may result in poor seed quality.

In controlling pests with chemicals, special care has to be taken with insect-pollinated crops, such as sunflower and onion. When insecticides are sprayed to control pests, the useful pollinators, such as bees, bumblebees and flies may suffer as well. This may result in poor seed set and thus poor seed yield.

Knowledge of seed-transmitted diseases is often not very well developed in communities that have otherwise developed a keen interest in experimentation with different varieties and cultivation techniques. The reason is probably that some diseases appear only long after crop establishment, so

Table 6.2: Major seed-transmitted diseases

	Disease	
CEREALS		
Maize	Black kernel blight (*Helminthosporium carbonum*) Diplodia rot (*Diplodia* spp.) Seedling blight (*Gibberella fujikuroi*)	Downy mildew (*Sclerophthora macrospora*) Bacterial leaf blight (*Erwinia stewartii*)
Pearl millet	Ergot (*Claviceps microcephala*) Grain smut (*Tolysporum pencillariae*)	Green ear (*Sclerosporella graminicola*)
Rice	Blast (*Pyricularia oryzae*) Brown spot (*Drechslera oryzae*)	Foot rot (*Gibberella fujikuroi*) Bacterial blight (*Xanthomonas oryzae*)
Sorghum	Anthracnose (*Colletotrichum graminicola*) Target spot (*Sphacelotheca sorghi*) Kernel smut (*Sphacelotheca reliana*) Head smut (*Gibberella zeae*)	Seed rot (*Fusarium moniliforme*) Bacterial streak (*Xanthomonas holcicola*) Bacterial stripe (*Pseudomonas andropogoni*)
Wheat	Smut (*Ustila go* spp.) Common bunt (*Tilletia caries*)	Head scab (*Fusarium* spp.) Spot blotch (*Helimithosporum sativum*)
PULSES		
Bean	Anthracnose (*Colletotrichum lindemuthianum*) Bacterial blight (*Xanthomonas phaseoli*)	Halo blight (*Pseudomonas phaseolicola*) Bean mosaic virus
Chickpea	Gram blight (*Mycosphaerella rabiei*) Wilt (*Macrophomina phaseoli*)	Fusarium wilt (*Fusarium orthoceras*)
Cowpea	Cowpea wilt (*Fusarium oxysporum*) Charcoal rot (*Macrophomina phaseoli*)	Cowpea mosaic virus Bacterial blight (*Xanthomonas vignicola*)
Groundnut	Rosette virus	Seed rot (*Fusarium* spp.)
Pigeon pea	Anthracnose (*Colletotrichum lindemuthianum*)	

	Disease	

OIL CROPS

Sesame	Bacterial leaf spot (*Pseudomonas sesami*)	
Soybean	Wildfire (*Pseudomonas tabaci*) Bacterial pustule (*Xanthomonas phaseoli*) Leaf and stem blight (*Colletotrichum truncatum*)	Seed rot (*Phomopsis* sp.) Stem rot (*Sclerotimia sclerotiorum*) Downey mildew (*Peronospora manshuria*)
Sunflower	Downy mildew (*Plasmopara halstedii*) White rot (*Sclerotinia sclerotiorum*)	Wilt (*Verticillium albo-atrum*)

VEGETABLES

Cabbage	Leaf spot (*Alternaria brassicae*) Black leg (*Phoma lingam*)	Black rot (*Xanthomonas campestris*)
Carrot	Bacterial blight (*Xanthomonas carotae*) Black rot (*Alternaria radicina*)	Cercospora blight (*Cercospora carotae*)
Cucumber	Angular leaf spot (*Pseudomonas lacrymans*) Anthracnose (*Colletotrichum lagenarium*)	Mosaic virus Scab (*Cladosporium cucumerinum*)
Eggplant	Blight (*Phomopsis vexans*)	
Lettuce	Anthracnose (*Marssona panattoniana*)	Mosaic virus
Onion	Purple blotch (*Alternaria porri*)	
Okra	Yellow vein virus	
Pepper	Anthracnose (*Gloeosporium piperatum*) Bacterial spot (*Xanthomonas vesicatoria*)	Blight (*Phytophthora capsici*)
Radish	Black leaf spot (*Alternaria raphini*) Black leg (*Phoma lingam*)	Black rot (*Xanthomonas campestris*)

(*continued over*)

Table 6.2: (cont.)

	Disease	
Tomato	Bacterial spot (*Xanthomonas vesicatoria*) Tobacco mosaic virus	Canker (*Corynebacterium michiganense*)

ROOTS AND TUBERS

Irish potato	Black rot (*Colletotrichum atramentarium*) Black leg (*Erwinia phytophthora*) Brown rot (*Xanthomonas solanacearum*) Late blight (*Phytophthora infestans*) Potato viruses	Ring rot (*Corynebacterium sepedonicum*) Rhizoctonia (*Rhizoctonia solani*) Scab (*Streptomyces scabies*) Verticillium wilt (*Fusarum oxysporum*)
Sweet potato	Black rot (*Encocomidiophora*) Internal cork virus	Scurf (*Monilochaetes infuscans*) Stem rot (*Fusarum oxysporum*)

that connecting it with seed quality is not easy. An example is smut in sorghum, which appears only after flowering.

The list in Table 6.2 is not complete. Some diseases appear in certain regions of the world only, others may in some areas not develop into major epidemics. It is beyond the scope of this book to present all the characteristics for identifying each individual disease, since these may be obtained from within the research or extension systems in most countries. It is, however, not always well known that such diseases may be seed transmitted.

There are basically two ways to avoid contamination of such diseases in seeds: disease management during crop production, and seed treatment.

Measures to control seed-transmitted diseases in the field
Disease management starts with the use of disease-free seed and fields. Crop rotation is an important tool in reducing the pressures of disease surviving in the soil. Many diseases may, however, be transmitted in different ways, e.g. by insects, wind or surviving in plant debris.

Disease control while the seed crop is in the field may involve the removal of diseased weeds and crop plants from the field (roguing), or selecting plants or portions of the field where plants look healthier to be harvested for seed. When pesticides are available to combat these particular diseases, they may be used on the whole field or applied particularly to

106

the portion that may be harvested for seed. Extensive or late spraying of fungicides is expensive, ineffective and a cause of environmental and health hazards.

Reducing disease by avoiding periods with high disease pressure is another important strategy. Early planting and the timely harvesting of the seed can be very effective. Another strategy is the production of seed in the off-season, i.e. the season that the crop is not widely grown in the area. This is normally only possible under irrigation. In some areas growing off-season is not possible because the crop is not suited to the season because of cool (winter) or hot (summer) weather or a different day length in which the crop may not flower at all. In particular areas it may be important to organize a period 'without host' for the disease (see Box 6.2), but this may not always be possible.

Box 6.2 Bean seed production in Kasese, Uganda

The area in western Uganda surrounding the town of Kasese is very well suited to bean production. The area consists of hills bordering a large hot plain. The largest part of this plain was turned into a national park. The Ministry of Agriculture has reclaimed part of the plain by developing an irrigation scheme. This scheme produces high-value crops like vegetables, and some field crops, such as beans and especially bean seed. Because of the supply of water and sunlight throughout the year, continuous cultivation is possible. This economic opportunity appears to be advantageous to various bean diseases as well, however, many of which are seed transmitted, such as BCMV (Bean Common Mosaic Virus).

Careful selection of virus-free foundation seed for the production of certified seed was not effective, because virus-carrying aphids were present throughout the year. The scheme was advised to introduce one 'break' season, i.e. three months with no bean cultivation in the irrigation scheme. The aphids had to feed on crops and weeds on which the BCMV cannot survive and re-infection of clean seed fields was minimized.

Other cultivation practices may reduce the disease incidence, such as mixed cropping. In fact, all measures that reduce disease incidence in a crop contribute to a better seed quality: they reduce the contamination of the seeds but also result in stronger plants, with better yields.

Seed treatments after harvest
Seed treatments, such as heating and the application of chemicals, are possible for a number of diseases. Heating may be used to control diseases located on or in the seed coat. It has to be done under extremely well controlled conditions in specialized installations, since the heat that kills the pathogens may also kill the seed itself. The temperature and the time of exposure have to be controlled very carefully.

Seed treatment chemicals have to be handled with great care. Selection and application of the chemical needs careful planning, and further handling of the seed, e.g. during manual planting, may pose a considerable health hazard (see Section 7.5).

There are local methods to treat seeds, but most of these are meant to kill insects. The control of virus diseases in seed is difficult. Viruses can generally be avoided only by selection in the field, combined with thorough weeding of possible alternative hosts in the weed population. Curative measures do not exist. Similarly, bacterial disease incidence cannot be reduced by chemical measures. Antibiotics are available, but should not be used for large-scale seed dressing. The best way to reduce bacterial disease incidence is long-term rotation and good seed selection in the field. The bacteria in Table 6.2 are the *Xanthomonas*, *Pseudomonas* and *Corynebacterium* diseases.

Fungus diseases can generally be controlled with seed treatment chemicals. These can be very effective, but cannot always be used in local seed systems because neither the chemical nor the right formulation are available, or simply because of cost considerations. The investment may be worth it, however, because seed treatment is often very cost effective compared with field sprays or crop losses in the following season (see Section 7.5) and less damaging for the environment.

6.4 Special groups of crops

6.4.1 Biennials

A group of crops where seed production requires special techniques is the biennial vegetables. These start to flower after an extended vegetative period and often require specific types of flower induction, consisting of cold treatment (vernalization) followed by long days. Examples are onion (and relatives), carrot, lettuce, radish, and cabbages (see also sections in Part D on these crops).

Seed production generally starts with normal crop production procedures, with the exception that the site has to be chosen very carefully. These crops are exotic to most tropical regions and seed production can often be realized only at higher altitudes or colder latitudes. Alternatively, vernalization may be done artificially, e.g. by putting onion bulbs in a cold store for a period of time before replanting. Sometimes, however, individual plants may be identified that will flower. Such 'early bolters' offer the opportunity of producing seed without going through complicated procedures. There is, however, a critical disadvantage in selecting these plants. Early bolting may be good for seed production, but it is a very negative characteristic for crop production. As soon as cabbages, radish, carrot or onions start to flower the vegetable product is wasted. When seed is produced using early bolters, the offspring are also likely to flower without 'warning'. This means that harvesting has to be done early (with a resultant loss in yield and quality), or the farmer will risk losing a considerable portion of the crop.

In lowland or medium-altitude tropical areas, local seed production of these crops is probably inadvisable. Areas above 1500m tend to be cool enough for seed production, but in many countries these areas are humid because of low cloud cover and thus not very suitable for seed production. Onions in particular are very sensitive to moisture both during flowering and in the seed ripening stage; radish and cabbage are more tolerant.

6.4.2 Pasture species

In most countries, cattle are allowed to feed on natural pastures or on pastures that are being improved by removing unwanted plant species only. Sown pastures are uncommon. One species which local farmers may plant is elephant grass (Napier) for zero grazing. In some countries, specialized fodder production is promoted using, for example, mucuna in open fields or *Centrosema* under trees.

There is, however, a great potential for boosting cattle production (beef and milk) through improving the pastures both in terms of total biomass production and of feed value. Such improvement includes the introduction of new species in the ley, especially legumes, or the use of selected varieties of common species, such as *Chloris* and *Panicum*.

Small-scale production of such seeds is being promoted in various countries, but this meets with a number of additional difficulties, compared to small-scale food-crop seed production. The main difference relates to the lack of local knowledge in this field. Farmers know how to produce food crops and the step from improving grain production to the selection of quality seed is relatively small. The promotion of pasture seed production includes teaching both the growing for seed and of cultivating a completely new crop. Secondly it requires the development of a totally new farming technique: the sowing of pastures. This means that the (local) market for improved pasture seeds is limited and unpredictable.

Important technical problems relate to the harvesting and cleaning of pasture seeds. The seeds of grasses are very small and those of legumes are difficult to pick and thresh. All species have their particular features, and it is beyond the scope of this book to go into too many details on these species. The International Livestock Research Institute (ILRI) is a good source of information.

6.4.3 Perennial crops

Among the most important food crops we find many fruit trees and estate crops. Many of these crops can be seed propagated, but vegetative propagation is widespread. The most important reason for vegetative multiplication is the possibility of maintaining and multiplying an outstanding plant. The resulting clone carries all the positive characteristics of the mother plant, whereas plants grown from seed are all different. The selection of relatively uniform varieties through a number of generations, which is common in annual crops, is not possible because it would take centuries to grow such generations from seed to fruit-bearing plants.

Many methods have been developed to propagate perennial crops vegetatively:

o The basic method is the use of rooted cuttings, i.e. to cut branches or twigs from the mother plant and stimulate the quick root development by using the right potting medium, protecting the leaves from drying and sometimes adding plant hormones.

o In some crops it is possible to obtain roots when a branch is bent down to the ground. The big advantage is that the leaves of that branch will not easily wither since they remain attached to the mother plant until the new roots take over the supply of moisture. The disadvantage is that the number of plants that can be produced in this way is limited.

o Another set of methods includes the grafting of stems, branches or 'eyes' into an already rooted plant (often a seedling). Apart from cloning a selected plant, this method offers the opportunity to select the plant for its production and quality potential only, whereas the rootstock may be particularly suited to an outstanding root development.

o In some crops where seed is used, we can still speak of vegetative propagation, (e.g. in Citrus species). The seed has the ability to grow plants not from the gamete (the combined tissue of the father and mother), but from maternal tissues only.

o Finally, micro-propagation under *in vitro* conditions is used for more and more crops on a commercial scale, especially for estate crops. These techniques are, however, beyond the scope of this book.

In all these methods it is extremely important to select the mother plants. The mother plant should carry all preferred characteristics regarding growth habit, production potential and product quality. This selection is particularly important since decisions affect the production for many years. Secondly, since it is vegetative propagation, special emphasis has to be given to diseases. Diseases that are present in the mother plant will be transferred to the new planting material without the barrier of a seed stage.

When all potential mother plants are diseased, tissue culture may be an effective way to obtain disease-free planting materials. In tissue culture a small part of a mother plant – in most cases a bud – is brought into a sterile culture. The tip of the bud does not normally carry the diseases that would be transferred through vegetative propagation. This initial stage is followed by a rapid multiplication phase, whereby large amounts of diseasefree material can be obtained. This micro-propagation requires some specialized equipment and expertise. When a disease is particularly serious, contracting out such specialized work may be a very effective way of assisting farmers.

Multiplication of perennial crops through seed normally involves the selection of particular sources, i.e. mother trees from which the progeny has proven useful. Such sources have to be protected. When mother trees become old their production level (e.g. fruits) may decline, but their value as seed sources may still be immense. Since most perennial plants are cross-fertilized, the trees surrounding the mother tree are pollen sources and should not be replaced.

In some crops the influence of the father may also be known, e.g. through hand pollination with a known source of pollen. Such hybrid seed production is becoming widespread in commercial plantings of estate crops such as coconut.

Whether perennial plants are multiplied vegetatively or by seed, in most cases this is done in a nursery. This allows the nurseryman to perform a selection according to plant type and health. Specialized manuals for nursery management exist.

7. Harvesting, processing and storage

Harvesting and post-harvest operations have an important impact on seed quality. This chapter introduces the general aspects of these operations, such as careful handling, seed drying, and the effects of temperature, moisture, and insects on seed storage. These aspects can be taken into account when assessing local methods of seed storage. Finally, chemical seed treatment is introduced.

7.1 Harvesting

Harvesting should be done in a timely manner to allow for quick drying of the seed, and to avoid important losses due to shattering or field infestation of storage insects (e.g. weevils in maize). Farmers often delay harvesting because of the peak labour demand at the end of the season, and because drying the crop on the plants reduces the need for drying floors and the labour associated with handling the seed there. Harvesting and threshing operations have to be done with great care to avoid damaging of the seed. Threshing when the seed is over-dried may lead to cracking of the seed and should be avoided, but also threshing wet seeds may cause internal damage, and result in subsequent germination and vigour problems. When seed is bagged, the bags should be clean, i.e. containing no remaining seeds from the previous harvest and no insects. Turning the bags inside-out is normally sufficient. Seed bags may be labelled in order to avoid mistakes.

7.2 Drying

Seed should be dried quickly, but high temperatures can cause damage. Sun-drying can normally be completed in a few days. For some crops special racks or cribs are used to improve the ventilation between, for example, maize ears or sesame plants, in order to quicken drying. When seed is dried on the floor, regular turning will improve the balanced drying of the seed lot and avoid mould growth at the bottom of the layer.

In humid climates, seed drying can be a serious problem. When harvesting cannot be done during a dry season, small-scale wood-fuelled dryers can be used. These require a considerable investment and experience is needed to avoid over-heating the seed. The effect of high temperature is most damaging when the moisture content of the seed is high.

7.3 Storage

High temperature and moisture are major enemies in seed storage. They affect the maintenance of seed quality in storage (see Boxes 7.1 and 7.2). Table 7.1 gives approximate periods that seeds of a number of crops can be stored under given seed moisture conditions. Additionally, high temperature and moisture favour the development of insects, bacteria and fungi.

Box 7.1 Harrington's rule for seed storage

One per cent increase in seed moisture content reduces the storage period. by half. (Or: one per cent reduction of seed moisture content doubles the safe storage period.)

Five degree Celsius increase in storage temperature reduces the storage period by half. (Or: reduction of the storage temperature by 5% doubles the safe storage period.)

These rules apply for most dry seeds in the most common range of seed moisture content and temperature.

Box 7.2 Example of seed viability after storage

Groundnut seeds are stored well at 6% moisture even at average temperatures. Seed of 95% initial viability will still germinate for 80% even after storage at 24 degrees for eight months.

When the seed contains 7% moisture, a germination of 80% is already reached after four months (126 days); the seed will germinate only for 55% at the onset of the season (eight months after harvesting).

Similarly, when the storage temperature is raised from 24 to 29 degrees (constant moisture content = 6%), 80% germination is reached after 121 days; when stored for eight months germination drops to 52%.

These figures were calculated with the SEEDLIFE computer program (see Note 1, Table 7.1).

Storage structures and practices should also protect the seed against the damage of rats and other rodents. Storage structures for food grains are often designed for the same purpose.

Temperature is normally difficult to control, apart from avoiding storing seed in direct sunlight or in hot places such as under a corrugated iron roof. Traditional storage structures, such as those using mud walls or underground spaces, are often well designed and provide efficient isolation to keep temperatures moderately low.

Ideally, airtight containers are used to store well dried seed. This is feasible for small quantities of vegetable seeds, but not for bulky field crop seeds. For vegetables, various glass jars are used, such as soda bottles

Table 7.1: Storage capacity of different crop seeds

Crop	Length of time of safe storage at 24°C (months)[1]		Maximum temperature for one year's storage (°C)[1]	
	RH[2] = 45%	RH = 75%	RH = 45%	RH = 75%
Barley	19	2	27	10
Pearl millet	13	2	24	8
Rice	13	3	24	11
Wheat	6	1	20	0
Rape	28	4	29	15
Pea	37	3	32	15
Bean	67	12	37	15
Cowpea	39	4	33	14
Broadbean	70	15	36	25
Groundnut	11	1	23	–
Soybean	17	2	26	10
Cabbage	23	4	30	17
Onion	13	1	23	10
Lettuce	35	3	30	15

Notes: 1. Calculated with the 'Seedlife' computer program, developed by CPRO-DLO, Wageningen, the Netherlands). The following conditions are assumed: germination before storage 90%, germination after storage 70%, no insect damage.
2. Relative Humidity.

Table 7.1 can be read as follows. When pearl millet seed with a germination percentage (viability) of 90% is stored during a rainy season (high humidity of the air, e.g. 75% and 24° C, second column) the viability will already have dropped to 70% within two months. If, however, the seeds can be packed in a moisture-proof container after thorough drying just after harvesting, resulting in a humidity of 45% inside the bag, the same seeds would be sufficiently viable even after 13 months of storage (first column). A similar result could also be obtained over one year storage by reducing the storage temperature to 8° C (fourth column), which is, however, practically very difficult under farmers' conditions.

sealed with candle wax. For larger quantities (a few kilograms) clean knapsack sprayer tanks can be used. In some countries, 50kg bags of laminated polythene/aluminium foil are available, others use multi-layer polythene-lined oil drums. Airtight containers also solve possible insect problems, because the insects suffocate as soon as the oxygen in the container is used up; it is therefore advisable to have the containers well filled, leaving little air in the container for the respiration of insects. Inert materials are sometimes added to fill the space between the seeds, so as to reduce the volume of air and restrict the movement of insects through the container. Sand or ashes can be used for this purpose. Projects in which small low-cost aluminium tanks were designed and produced locally have been successful for improved maize seed storage in Central America and other places.

It is extremely important that seed in airtight containers is dried very well before the container is closed, especially when the storage season is warm and humid. Some respiration will occur, thus increasing the relative humidity (RH) in the container. This problem can be reduced by introducing layers of fresh charcoal in the container, separated from the seed by newspapers. The charcoal absorbs the humidity. In most cases, however, seeds have to be stored in normal (gunny) bags or in bulk. In this case, the storage conditions are very important: cool, dry and without insects or rodents.

The storage of the seed has to be safe, i.e. safe from thieves and from fire and other calamities. Grain stores are generally well protected from rain and rodents (rat traps at the poles under the store). Hollowed-out gourds are inverted over hanging seed ears to provide protection against rodents. In Mali, small amounts of cucurbit seed are protected by mixing them with cattle dung or mud and plastered as a cake against a mud wall, under a roof (Wright et al., 1994).

Seed is also often stored in the house for even better protection and for safety. An exceptional case is the hidden underground stores used in some parts of Ethiopia. They have proven safe even after the displacement of farmers during civil unrest. The storage of seed ears in the kitchen hanging in the smoke of the fire is not only a safe place, but also keeps the seed dry and reduces insect and disease damage.

Mixing beans with ash is reported to reduce damage by bruchids and other insects. Ash damages the cuticle of the insects, causing them to dehydrate. Ash should be added in sufficient quantities; 25–50 per cent by volume is recommended. Lime and diatomaceous earths added to the ash are suggested to improve the effectiveness of the protection (Greve, 1983). Vegetable oils, such as for example soya oil, can be used as a dressing to reduce insect damage. Damage by bruchid and acanthoscelides is reduced by mixing 5–10 ml of vegetable oil with 1 kg of beans (Greve, 1983). In Northern Ghana, cowpeas are mixed with shea butter oil and left in the sun as protection against bruchids (Wright et al., 1984).

Various plants and plant extracts are used as repellents in different parts of the world, such as crushed seeds or leaves of neem, eucalyptus or lantana. It must be borne in mind, however, that natural substances may be as toxic as chemical biocides and should also be treated with care. For insect control, the application of chemicals can be very effective (Section 7.5).

7.4 Seed cleaning

Seed cleaning has a dual purpose: it removes non-crop seed materials from the harvested material, such as straw, stones and weed seeds, thus reducing the bulk that has to be stored; and it also allows the selection of seeds according to physical characteristics, such as size, shape, density and colour, thus removing shrivelled and small seeds, and increasing the seed quality.

Seed cleaning is for most crops basically the same operation as the cleaning of food grain for consumption, and local methods used for food grains are in such cases quite well suited for seed cleaning. Such methods include winnowing, sieving and hand-picking. Winnowing removes the light particles like straw and dust, and it can be used to remove seeds with a low density (low weight per volume: empty or 'soft' seeds). Sieving selects the seed on shape and size. Picking the seed by hand is used to remove diseased and discoloured seeds.

The formal seed sector invests large amounts of money in seed cleaning machinery. The most common machines include the clipper, which combines sieves and a fan, blowing out light particles. An indented cylinder is commonly attached to the clipper, and is used for sizing the seed. Optional extras include a gravity table for removing stones and light seeds, a colour sorter to remove off-coloured seeds and various specialized machines such as a needle separator to remove insect-damaged legume seeds, a magnetic roll to remove dodder (*Cuscuta*) from alfalfa seeds, etc.

These machines are efficient, but often do not give a better result than hand-operated winnowing, sieving and picking. When (women) farmers pick their bean seeds just before planting, they can remove disease-stained seeds much better than any electronic colour sorter. The limitation of hand-sorting seeds is, however, the time it takes.

Relatively small seed-cleaning machines are available (0.5t per hour). The necessary investment and their requirement of electricity make such machines difficult to operate in many places. Such investment should be considered very cautiously since they are needed for relatively small quantities (compared to grain), and for a limited period during the year. Often it is better for farmers or farmer groups to approach an existing plant in the formal sector, and ask for contract cleaning of farm-produced seed.

7.5 Chemical seed treatment

In some cases, crop dressing with chemicals can be useful, to fight storage and field insects and seed-borne and seedling diseases. Seed treatment can, on the other hand, be a most effective and relatively environment-friendly means of pest and disease control. Seed treatment uses small quantities of chemical compared to field sprays and can be very effective beyond the germination of the seed. Beans treated with endosulfan appeared to resist the attacks of bean fly in Central Africa, which cause great yield reductions. Vitavax is a systemic fungicide, i.e. it is transported throughout the germinating plant, and is very effective in controlling various smut diseases in cereals which are difficult to control through field sprays. Vegetable seeds can be mixed with carbaryl against insects and stored in airtight containers. Actellic (pirimiphos methyl) is a widely available insecticide in Africa which can be applied to stored seeds.

Chemicals in local seed systems should be used with extreme care. This includes both the choice of the chemical and the application method.

Commercial seed treatment chemicals are often very toxic. They are developed for use in advanced seed-processing plants, and meant for mechanized farming. In those situations the contact with workers and farmers is minimized. Farmers receive treated seeds in closed bags which are emptied into the hoppers of tractor-propelled planters. This is why toxicity standards for seed-treatment chemicals are generally less strict compared to field-sprayed biocides. In less-mechanized farming, however, both the treatment of the seed and the planting by hand increase the contact, and the chances of endangering persons who have not been trained in safe biocide handling (e.g. knowing not to smoke or eat while or just after handling poisonous chemicals, wearing protective clothing and masks, and washing after using biocides, etc.).

Seed dressings can be applied in various ways:

○ spreading the seed on a tarpaulin, sprinkling the chemical and turning the seed with a shovel;
○ using a tilted oil drum, supported on a stand and turned with a handle (see Figure 7.1);
○ using a commercial concrete mixer.

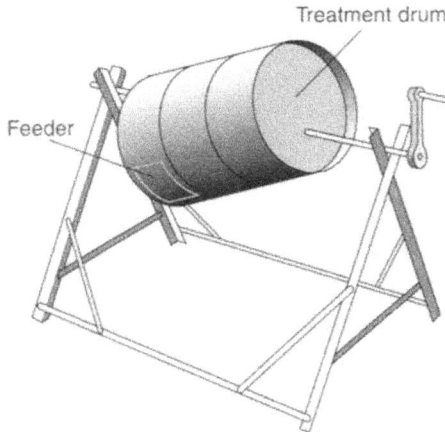

Figure 7.1

Specially designed seed-treatment machines usually achieve a more even coverage and therefore a greater cost-effective use of the chemical. They also allow for a safer application of these toxic chemicals. The above methods are, however, quite effective as well.

With regard to health hazards, the formulation of the chemical is important: the use of dusts will lead to inhalation of the chemical (more so with the shovel-method than with the others). Liquid and slurry dressings are generally safer but may require additional drying of the seed after treatment.

117

An indication of the toxicity of the most common chemicals is presented in Table 7.2.

Table 7.2: Some chemicals for seed treatment and their general characteristics

	Acute toxicity (LD50 in mg/kg)*	Persistence	Used for	Comments
INSECTICIDES				
Lindane	high (125)	long	only seeds	
Carbofuran	very high	very long	caution	3
Endosulfan	very high	very long	caution	3
Malathion	low (2800)	moderate	seed/grain	1,2
Actellic	low (2050)	good	seed/grain	2,5
Pyrethrum	low (1500)	low	many purposes	4
FUNGICIDES				
Thiram	moderate (780)	low	seed/fruit	6
Captan	low (9000)	low	seed/fruit	6
Carboxin	low (3820)	moderate	seed	7

Notes:
* LD50 is the dosage of the chemical which is lethal to 50% of the rats to which the chemical is administered (in mg chemical per kg weight of the rats). This is a good indication for acute toxicity in humans.

(1) Resistances have built up: not effective any more in all areas.
(2) Not effective against the grain borers.
(3) So toxic that their use should not be promoted.
(4) Pyrethrum is a natural insecticide, which has been copied chemically and marketed under many names, mainly ending with '-thrin'. It is mainly used in mixtures with other chemicals.
(5) Actellic is a trade name of Pirimiphos-methyl.
(6) Very widely used fungicides for control of a wide range of seed or soil-borne fungi.
(7) Systemic fungicide used to control smuts in cereals and Rhizoctonia. Vitaflow is a combination for seed treatment of carboxin and thiram which controls a very wide range of fungus diseases.

In very advanced seed systems various non-chemical seed treatments have been developed. Very small or irregularly shaped seeds are pelleted, i.e. a fine clay-like substance is pasted around the seed. Such pelleted seeds, e.g. of lettuce or beet, are larger and evenly shaped which allows precision planting. Other techniques have been developed to increase uniform germination. Intensive horticulture requires very uniform crops, and very even emergence of the seeds. Pre-germinating and subsequent re-drying of seeds is called priming. Priming has to be done under very well controlled conditions, and the handling of primed seeds requires special care. Pelleted seeds are exported to developing countries; primed seeds rarely.

8. Seed and variety selection

Selection is an important activity in a seed-improvement programme. It is, however, often assumed that variety maintenance and crop improvement are specialized activities that can be performed only by trained breeders. This chapter presents some basic background and techniques which may be used in participatory approaches to plant breeding introduced in Chapter 2 and specified in Chapter 12. The time of selection and the mating system of the crop are important aspects.

8.1 Options for selection

Selection is an important aspect of seed production. Selection is carried out in order to:

○ improve the vigour of the seed by selecting well-developed plants and plump seeds only (physiological and analytical quality);
○ reduce disease incidence by discarding obviously diseased plants or seeds (sanitary quality);
○ maintain the genetic quality of the variety;
○ adapt the variety continually to changing growing conditions; and
○ obtain better varieties.

There are different selection methods, realized during different phases of seed production. Going from planting time backwards in time these are:

(1) Selecting before sowing healthy and true-to-type seeds from the bulk of the stored grain, i.e. seeds that resemble the seeds of the mother crop and that do not show obvious disease symptoms.
(2) Selection after harvesting, but before threshing and storage. This is a common method in maize and sorghum, where the best-looking ears are kept separately for seed. An advantage is that for small amounts of selected seed drying and storage conditions can be given more attention.
(3) Selection during or at the end of the season of a particular field or part of the field that performs well. This portion is harvested separately for seed. The advantage is that seed from well-developed plants with fewer disease symptoms are expected to have better vigour and higher sanitary quality.
(4) Picking individual plants just before the harvesting of the whole field is done. The advantage over method (3) is that this can include some selection of the genetic composition of the variety.

(5) Marking particularly healthy and good-looking individual plants during the season, e.g. by tying a ribbon around the stem of the selected plant, and harvesting separately (positive mass selection). At the same time, off-type or diseased plants may be removed from the field (negative mass selection). Compared with method (4) this allows for selection of characteristics that are no longer visible at the end of the season, e.g. susceptibility to leaf diseases.

(6) Selection of a field for seed production, separate from the crop production field, taking into account some isolation distance. A combination of this practice with the roguing of off-type and diseased plants during the season can increase the selection pressure.

(7) Performing specialized selection procedures to maintain or purify the variety (see following section on variety maintenance).

(8) Selection of a totally different environment for seed production, e.g. seed potato production at a higher altitude to reduce disease pressures while potato crop production is maintained in the valleys.

The options (1) to (8) are listed in order of specialization and, with regard to genetic selection, in order of efficiency. There is a break between options (5) and (6): in options (1) to (5) seed production is a part of the crop production process, while options (6) to (8) relate to a specialized seed production, separate from crop cultivation.

In the first option the selection does not necessarily relate to the characteristics of the plant and thus bears the risk of genetic degeneration of the variety. For example: when some trailing beans are mixed with seed of a pure bush-type variety with similar seed characteristics, the quantity of trailing plants will increase with time because the trailing plants produce more pods. Seed selection before planting can, however, be important in sorting out obviously diseased seeds, e.g. brown-spotted bean seeds with fungus or virus disease, and poorly developed seeds.

The options (6), (7) or (8) may be useful in very special cases, i.e. where there is a great need for uniform varieties or where disease pressures are high. In large-scale seed production this is by far the best option, but in common agricultural practice this is done only in very specific situations. In common farming, farmers generally perform selection methods (1), (2), (3), or (4), or a combination of them. There are cases, however, where some farmers cultivate in conditions particularly suited to seed production, or where particular farmers or farmer groups have a special interest in producing superior-quality seeds, and in these cases have developed more sophisticated seed production and selection practices. In many situations options (3) and (5) offer interesting opportunities to improve seed quality.

An effective refinement of options (4) and (5) can be the so-called 'grid selection'. In this method plants are selected in relation to their neighbouring plants. This means that plants that look particularly good compared to their neighbours should be selected instead of all the plants from a possibly more fertile part of the field. A good farmer will select outstanding plants from all corners of his field by comparing each plant with its neighbours.

Seed cleaning and the removal of diseased plants from the field is also a kind of selection. Even though such selection on non-genetic seed quality is important, selection is more commonly associated with variety main-tenance or improvement, which are dealt with in Sections 8.2 and 8.3. It is important to realize, however, that any selection on non-genetic seed quality may have (unintentional) effects on the genetic quality of the seed.

An example is the selection of plump cereal seeds from the bulk of a seed lot of e.g. wheat.

○ The seed may have grown on a plant which, by chance, was standing on a good site, e.g. on top of a large amount of cow dung, in a small depres-sion of the field where more water was available, or on a spot where neighbouring plants were destroyed by wild pigs. Selection of such plump seeds may not result in a marked genetic improvement.
○ The seed may have grown on a very vigorous plant, which is particularly well adapted to the conditions of that particular year. The plant may be resistant to a disease which caused other plants to wither and produce shrivelled seeds. The seed will produce a plant with the same characteris-tics that will prove positive when the conditions of the following year are similar. There may thus be a positive effect.
○ The seed may have developed on a plant where most ears had been eaten by a cow at emergence. This means that the plant would have used all its nutrients on this one remaining ear. The plants grown from the seed may not be particularly good or bad.
○ The seed may have developed on a plant which matured later or which pushed away all its neighbouring plants. This could be genetic and the resulting plants from those seeds may even be less adapted than the parent crop, because the extended maturity period may cause low yields when the following season is short.

It is thus not always easy to predict the effects of selection, and farmers' selection methods can sometimes be detrimental (see Box 8.1).

8.2 Variety maintenance

8.2.1 The maintenance of diversity

Farmers generally select in their fields and in their seed stores. This selection can run counter to natural selection, which introduces weedy characteristics in the crop such as shattering seeds, weedy plant architecture and other characteristics that may be positive for plant survival, but negative for crop production. Variety maintenance is thus an important aspect of seed supply.

Box 8.1 The effects of selection

Farmers often possess very well-developed methods for selecting their seed. It is not wise, however, to assume that such farmers' methods are always most effective. Clear examples are french bean, vegetable, and hunger crops in many parts of the world.

o Early in the season, when vegetable prices are high, farmers are tempted to sell their french beans. Only at the end of the season do they start bothering about seed for the following season. Some pods which develop late are left on the plant to produce seeds. Such plants may have accumulated all kinds of diseases, some of which may be seed transmitted. Moreover, the last seeds on a plant are commonly smaller, producing less vigorous plants.

o Tomato and melon growers have been observed to sell their best-looking fruits in the vegetable market to fetch a premium price. Oddly shaped fruits are eaten or used for seed. In case the malformation of the fruit has a genetic background, such methods increase the chances of producing more poor-quality fruits the next season.

o Farmers often have particularly early-maturing varieties that provide food for the hunger season, i.e. the period before the harvesting of the main food crops. Such early varieties are under severe pressure to be used for food rather than for seed. In early-maturing maize crops, the earliest cobs are commonly picked for cooking, roasting or selling. This is likely to result in a gradual delay of the maturity period of this early-maturing variety. Similarly, early-maturing varieties of bean such as *mesi moja* in Kenya ('one month') and '*saca pobre*' in Costa Rica may be lost completely, when the temptation is to consume or sell the whole early crop.

The maintenance of local varieties (landraces) is a dynamic process whereby the farmer often selects on the basis of a particular diversity within the variety which is characteristic for that particular landrace. This is different from modern plant breeders, who have an idiotype in mind, an ideal plant type that should be selected; a strategy which leads to uniformity.

In practice, farmers do not maintain the variety in a strict sense. They maintain the major features of the variety, but at the same time they can continually adapt the variety to changing conditions, e.g. the gradual reduction of soil fertility, or to specific changes in the market. The genetic diversity which is present in landraces gives the farmer the opportunity to respond to such changes. Selection within a genetically uniform variety does not in principle lead to marked changes.

Cereal farmers often sample the diversity of their landrace when they pick individual ears for seed before harvesting the food crop. In the process they may discard the obviously weak and diseased plants, thus applying some selection pressure. The basis is, however, to maintain the diversity within the variety. They rarely select only the very best-looking and very similar ears.

Another example is women bean farmers in Rwanda, who have been observed to mix different coloured seeds intentionally to arrive at well-balanced varietal mixtures for planting in different plots of their farm: mixtures for good soils, for shaded plots, etc. They could easily select single-coloured seeds which are likely to result in more uniform crops, but the diversity offers a buffer to various possible uncertainties during the coming season, such as disease outbreaks, drought, etc.

Maize farmers maintain their varieties by choosing the average ears, and not the largest ones, which if they did so would lead to changes, not only in ear size, but also other characteristics (Box 8.2).

Box 8.2 Poor selection in formal systems: KWCA maize in Uganda

Kawanda Composite maize in Uganda became very popular upon its release in the early 1970s. Poor selection over a number of years caused considerable changes in the variety that made it far less adapted to local conditions.

Selection method (2) was applied over a number of generations: large ears from a specially planted field were selected for maintaining the variety. Plant characteristics were not taken into account. The result was that after a number of years the average plant height and the maturity period had increased and the number of ears had decreased to one.

That the crop had changed in appearance can easily be explained by the selection method. Plants with large, better-looking cobs generally have one cob per plant, whereas plants with two slightly smaller cobs may have a higher yield. Similarly, plants with large cobs are very competitive: they must have intercepted more sunlight than their neighbouring plants: they were taller, and they remained green longer.

The result was that the increased maturity period caused problems with growing two crops per year and with planning the harvesting during a dry season. Tallness caused lodging problems at the end of the season. One cob per plant was considered inferior to two cobs.

It took a number of years of very specialized selection procedures to re-select KWCA to look more like the original variety.

The selection of landraces is more effective using methods (1) to (5) (Section 8.1). Methods (6) and (7) are particularly aimed at strict selection, i.e. increasing the uniformity of a too-heterogeneous variety. This should be done with great caution, clearly defining objectives and including regular adaptation and yield stability tests. Too strong selection leads to genetic narrowing down of the variety, which may reduce yield stability and reduce yield potential due to inbreeding depression (in cross-fertilizing crops).

Supporting farmers in the maintenance of heterogeneous varieties therefore has to be done with caution. It involves creating awareness about which types should be considered off-types (i.e. not belonging to the landrace), and removing such plants from the field, preferably before flowering.

Selecting within a variety to adapt it intentionally to changing conditions or new needs is discussed under the heading 'crop improvement' (Section 8.3).

8.2.2 The maintenance of uniformity

Selection in a modern, uniform variety is different. The aim in most cases is to avoid genetic degeneration, i.e. to maintain or to re-select the original variety. Regular selection is necessary to avoid the accumulation of off-types, which may not be optimally adapted to the conditions. Simple selection includes roguing of obviously off-type plants in the field, preferably before flowering of the crop, or during several rounds both before and after flowering (negative mass selection). Alternatively, a positive mass selection picks the best (true-to-type) plants out of a field, the seeds of which are then further multiplied. With relatively pure varieties of self- and cross-fertilizing crops this is effective and sufficient to keep a variety sufficiently pure. Positive mass selection for uniformity in modern varieties of cross-fertilized crops is described in the section on cross-fertilizing crops, Section 8.2.5 (see also Figure 8.1).

When a good variety has become mixed during subsequent seasons of reproduction, the application of method (6) may be necessary to re-select the original variety. This may look like a very laborious and specialized task, but it can be a very effective way to assist farmers to improve their seed. Basic selection schemes are presented here in order to guide such specialist selection. A distinction has to be made between vegetatively propagated, self-fertilizing, semi-cross-fertilizing, and cross-fertilizing crops. The last two methods need separate fields to avoid unwanted cross-pollination and close attention has to be paid to avoid inbreeding depression in cross-fertilized crops.

8.2.3 Self-fertilizing crops

When only little heterogeneity is observed in a uniform variety, simple mass selection can be used to maintain the variety. Mass selection can be done by removing the off-type plants (negative mass selection), or by positively selecting the preferred plants (positive mass selection).

When a uniform variety of a self-fertilizing crop shows considerable variation in the field a so-called pure line, or ear-to-row selection can be performed (Figure 8.1). The last name clearly expresses the method:

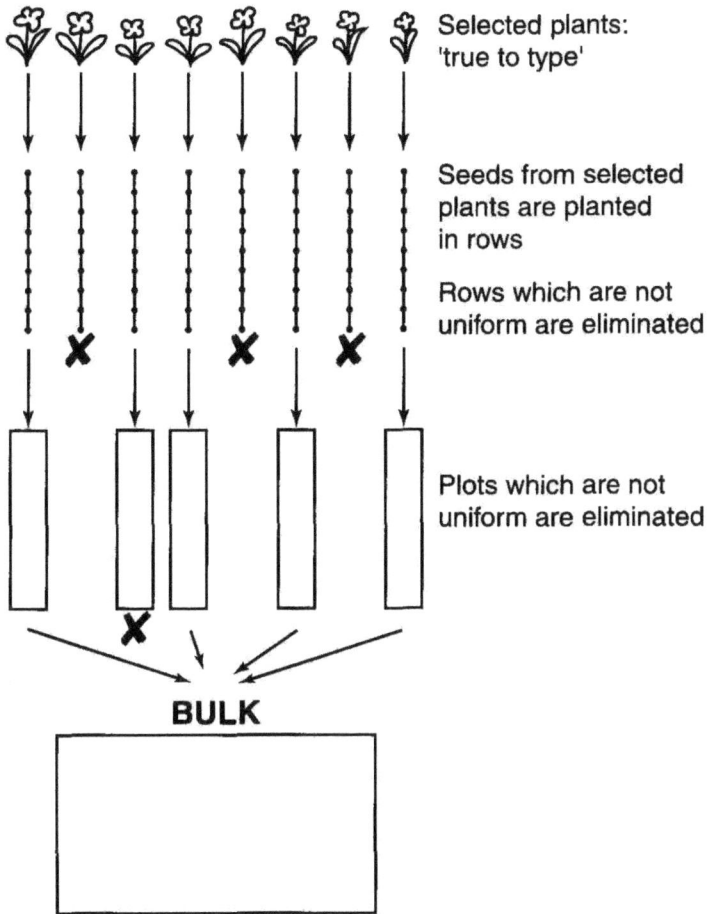

Figure 8.1 Pure line, or ear-to-row selection in self-fertilizing crops. Seeds from individual plants are planted in separate rows. Only from uniform-looking rows with true-to-type plants are the seeds harvested and bulked for the next planting.

(1) At least 100 true-to-type ears (in cereals like wheat or finger millet) or plants (in pulses like beans or chickpeas) are selected and harvested separately.
(2) The seeds from each plant are then planted together in separate rows or small plots.
(3) When the mother plant was genetically pure the row planted from the ear or plant will be very uniform. If the mother plant was not pure, the row will show a segregation, i.e. there will appear clear differences between the plants within one plot. In this case the whole row should be eliminated.
(4) Only rows that are uniform and definitely true to the variety are harvested for seed.

(5) The seeds from the different uniform-looking rows can be bulked.
(6) If an extra cycle of selection is still needed for more uniformity, the seeds from the different uniform rows are kept separately. They can be planted in blocks in the following season. The blocks will be larger than in the preceding season, allowing more precise evaluation. Non-uniform blocks are eliminated.
(7) Selected blocks that are similar are blended and multiplied for distribution to the farmers.

It is strongly advised not to keep a single uniform row or block for further multiplication. This particular row may have invisible faults, such as a poor disease resistance for a disease that is not very prominent during that particular season. It is better to select 10 rows or more which look similar. Moreover, this favours a more rapid multiplication.

When different modern varieties of a particular crop have to be maintained in the same field, some rows have to be planted around each selection-block separating the different varieties. These rows should not be harvested for seed, because some cross-fertilization or mechanical admixture may occur even in self-fertilizing crops.

8.2.4 Vegetatively propagated crops
Variety maintenance of vegetatively propagated crops is easy from a genetic point of view. Single plant selection or single plant selection combined with 'ear-to-row' multiplication will eliminate diseased and off-types very effectively. The main worry of maintaining varieties (clones) of vegetatively propagated crops is to keep the stock free from diseases. Very strict selection in rows for disease-free plants can be effective, assuming that control measures are taken to avoid the spread of diseases from one plant to the next within the selection field. This can be done by planting other crops between the lines (e.g. soybean in a bean selection field), or by early sprays with fungicide against fungus diseases. Regular spraying of systemic insecticides against the spread of insect-transmitted virus diseases can be useful when applied in large areas.

A variety may be infested completely with virus diseases. When healthy plants cannot be found, it may be possible in a research station to eliminate the disease through tissue culture. This is a costly exercise which is very useful only if there can be some guarantee that re-infection with the disease in the field can be avoided. If the likelihood of re-infection is high, a multiplication scheme may be considered in a 'clean' environment on a large scale. This can then be distributed to replace all infested materials in a particular area (e.g. virus-infested cassava).

8.2.5 Cross-fertilizing crops
Maintenance selection of cross-fertilizing crop varieties is more complicated than the maintenance of self-fertilizing varieties. The main difference

is that such varieties may suffer from 'inbreeding depression', when the variety becomes too uniform. Selecting a small number of plants may also result in genetic drift, i.e. a gradual shift in some characteristics of the variety. This means that selection does have its limits: too strong selection will result in a gradual reduction in yield.

A second problem with cross-fertilizing crops is that good-looking plants may have been fertilized by very poor plants. The selection of seeds from such good-looking plants thus multiplies the good characteristics of the mother plant, but also the poor characteristics of the pollinators.

The most simple procedure is mass selection: to remove off-types (negative mass selection, Figure 8.2) or to pick the best plants compared with the neighbouring ones (positive mass selection, Figure 8.3). In the latter case one should guard against selecting too few plants to avoid the risk of inbreeding depression or drift, which would render the whole activity counterproductive. Grid selection involves selecting the best plants from different parts of the field based on their good characteristics compared with neighbouring plants (see Section 8.1 and Figure 8.4).

A more advanced maintenance procedure is the following adapted ear-to row selection (Figure 8.1):

(1) Select at least 200–500 good-looking ears (maize, pearl millet) or heads (sunflower), i.e. those that are well developed, healthy and having all the typical characteristics of the variety, though not necessarily the biggest.
(2) Plant rows with the seeds of each plant. These rows may consist of 10 to 50 plants (the so-called 'half sibs', i.e. having the same and known 'mother') depending on the available field size.
(3) Remove the poor-looking rows, preferably before flowering.

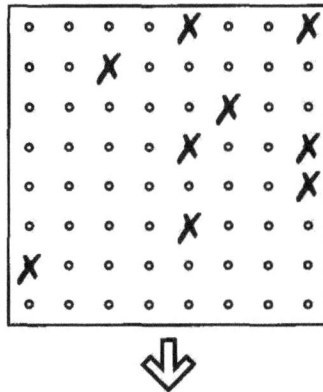

Figure 8.2 Negative mass selection. Off-type or bad-looking plants are eliminated; the harvest from the rest of the plants can be used for seed.

127

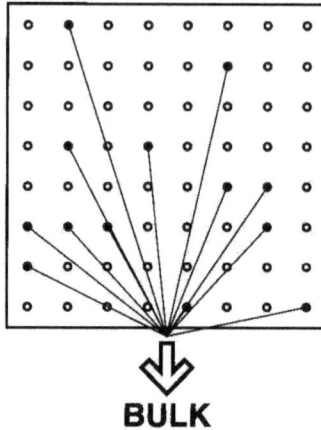

Figure 8.3   Positive mass selection. Seeds from the best plants only are selected and used for next season's planting.

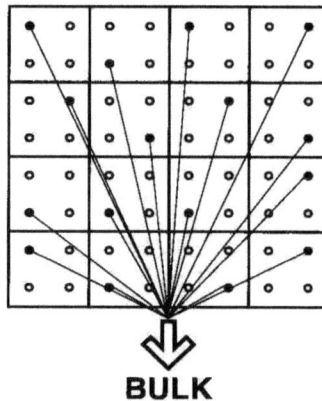

Figure 8.4   Stratified mass selection or grid selection. Seed is selected from plants distributed equally over the field. This method reduces the risk that differences in field conditions (soil fertility, irrigation) results in the selection of plants from only one side of the field.

(4) Remove the most irregular rows, preferably before flowering, and harvest the other rows and bulk the seed.
(5) Select the best plants or ears within the good rows to start a new selection cycle.

This selection is quite 'soft' and will be effective when executed for a number of seasons.

When a variety has been 'contaminated' because of many years of production without selection, or a problem with isolation, the above procedure may be refined with the following methods: (1) remnant seed method; (2) full sib selection; and (3) uprooting method.

Remnant seed method
This method is an ear-to-row selection whereby part of every ear or plant is kept in store. This is done because the plants of the selected lines in the field may be pollinated by neighbouring (non-selected) lines (Fig. 8.5), which considerably reduces the selection efficiency.

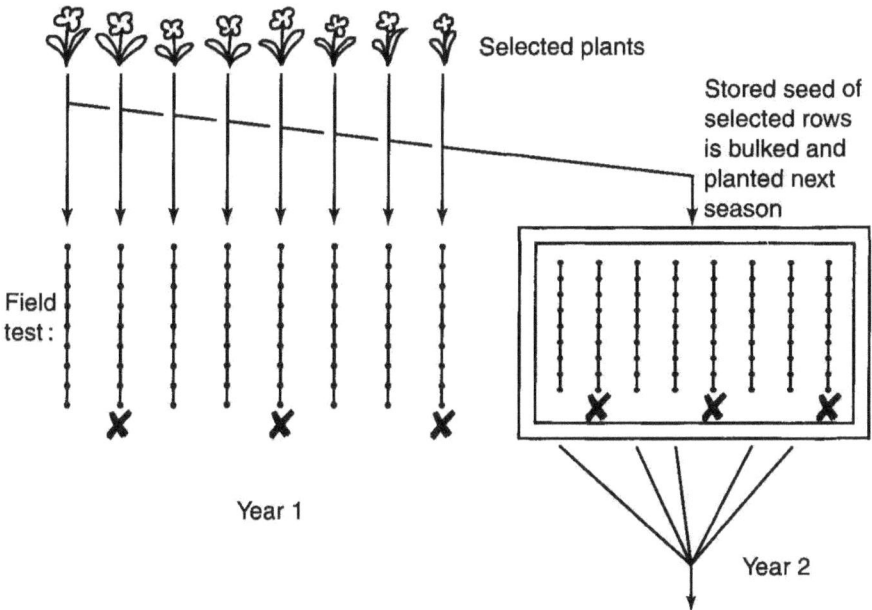

Figure 8.5   Remnant seed method for (semi-)cross fertilizers. Seed is harvested from selected healthy and true-to-type plants; seeds of each individual plant are kept separately. Only half the amount of seed harvested off each plant is sown in rows. When a row looks uniform, the seed that had been kept behind is bulked with the retained seed from other rows.

(1) When the rows are planted (at least 200), some seed of each plant/ear/cob is kept in properly labelled bags and stored (seeds from each individual plant in a separate bag with a label that corresponds with the label of the same seed that is planted in the field).
(2) The rows may remain in the field without removing off-types. Poor plants are labelled during several selection rounds during the season. Only the best lines (at least 50) are selected. The remnant seed is taken from the store, bulked and planted the following season. If fewer than 50 lines are selected there is a serious risk of in-breeding and thus reduction in yield.

Full sib selection

Instead of picking good-looking ears or heads, two good-looking plants may be artificially crossed. This is relatively easy with maize (see Figure 12.5), but very difficult in many other crops. The crossed ears (from 200 such crossings) then give the seed for the rows for the selection (with or without the remnant seed method). When a variety has become relatively pure i.e. relatively homogeneous rows or plots, these selected lines may then be planted in the middle of a field, surrounded by plants from remaining seed of the selected heads. This is often proposed for maize.

(1) Enough seed of the crossed parents is used to plant a row (5 or 10m long). These seeds are full sibs, i.e. sharing the same mother and father.
(2) At least 200 of such rows are planted in a block.
(3) Remaining seed of this full sib crosses is mixed and planted around the block of lines.
(4) A first selection round is made (rows that are insufficiently true to type or too heterogeneous are removed before flowering).
(5) In order to avoid inbreeding, plants in the best rows are detasselled, to allow them to be fertilized by plants from the surrounding field.
(6) Seed is collected from the (detasselled) rows that appear sufficiently true to the variety also after flowering. The seed is thus produced on strictly selected mother plants with a less strictly selected pollen cloud from the surrounding borders.

CIMMYT has developed a protocol for this kind of field lay-out for maize.

The selection of full sib 'rows' should be done on the major characteristics of the variety and on uniformity. These should not be selected for yield because heterosis may 'blur' the observation. Full sib selection is also used by breeders to obtain inbred lines and to assess combining ability.

Uprooting method

In cross-pollinated biennial root crops like radish and carrot, seed harvested from at least 200 good-looking plants can be planted in rows. The plants are uprooted and selected on root shape and size. The best roots (at least 100) of the best rows are then replanted in a block, surrounded by the second choice roots (at least 1000) of the good rows. All other roots, including very beautiful ones from variable rows can be eaten.

This block is allowed to flower and set seed. Seed harvested on the 'inner block' is used for further selection. Seed harvested from the 'second choice' plants can be used directly as a 'first-season selected seed'. This method can also be used for crops like cabbage, whereby plants can be transplanted as soon as the head characteristics can be observed. Also in this system, a sufficient number of plants has to be selected to avoid inbreeding.

The maintenance of different varieties in the same field is virtually impossible for cross-fertilizing crops. The isolation distance between the fields has to be large in order to avoid pollen from one variety contaminating the other. Planting tall crops between two varieties may reduce this risk in wind-pollinated crops (e.g. tall elephant grass surrounding maize selection plots), but not in insect-pollinated crops (sunflower, radish, cabbage).

In some conditions it is possible to stagger the planting of different varieties: planting the different varieties at different times prevents them flowering simultaneously. If conditions permit, however, it is much safer to concentrate on one variety in one year, and on another in the following year.

8.2.6 Semi-cross-fertilizing crops

The selection of semi-cross-fertilizing crops does have to take the crossing behaviour into account, but these crops generally do not suffer from inbreeding depression. This means that maintenance selection looks like the method used for self-fertilizing crops (ear-to-row) when the important characters can be observed before flowering, i.e. before poor plants can contaminate the selected ones.

When important characteristics cannot be observed before harvesting (e.g. seed colour in sesame) and selected well-performing plants may have been fertilized with pollen of less-performing ones, the remnant seed method presented above for cross-pollinators can be followed to increase efficiency. There is no need to observe the relatively high minimum number of selected plants in this type of crop, because inbreeding depression is not a danger. As in the situation in self-fertilized crops, however, selection of less than 10 plants should be avoided.

8.3 Crop improvement

8.3.1 General

The terms 'crop improvement' and 'plant breeding' are usually reserved for formal variety development activities at research stations and private seed companies. However, farmer-selection can also aim at crop improvement.
There are basically three methods of crop improvement:

○ replacing the existing variety with a new one that fits the farming system;
○ selecting within an existing variety until the features have significantly changed; and
○ creating variation through crossing plants, after which selection results in a new variety.

Local crop improvement as carried out by farmers may include any of these three options. It may involve the testing of unknown varieties under local conditions, the selection within varieties of off-type plants, or for their best characteristics. The third option is less common, but conscious hybridization between varieties occurs and consciously stimulating natural introgression by planting different types close to each other is even more common.

The knowledge of farmers who apply such methods is valuable, and joint efforts on crop improvement involving both farmers and trained breeders could increase efficiency where trained plant breeders are unable to cater for the diverse needs of farmers.

Section 2.2 introduced participatory crop improvement as a way to support communities in their efforts to develop better adapted varieties. This section will introduce some of the choices that can be made in setting up such activities.

Since the process is necessarily participatory, this book cannot 'prescribe'. The following section is meant to illustrate the wide variety of options for participatory PVS and PPB.

8.3.2 Participation from the start

When starting with a programme, the first step is consultative priority setting. Through participatory rural appraisals both breeders and farmers get to know the real priorities with regard to varietal characteristics. Where farmers express their priorities, breeders may know sources of preferred characteristics and the possibilities to select for them (heritability).

Proper analysis of the positive features and the limitations of the existing varieties is the second step. It is assumed that the existing varieties may not be optimal, but at least are well adapted to local conditions. In this process the varieties of nearby communities may also be included. This is an opposite approach to most national breeding programmes, which concentrate on materials obtained from abroad (often from international research institutes) and aim at adapting them to local conditions. The analysis of local materials will yield a list of less preferred characteristics (defaults), which may be worked on in the following stages.

When very small defaults are identified, it may be possible to select within the existing materials in order to remove the poor characteristic. There is evidence that the yield of local varieties of cross-fertilized crops can be increased by up to 15 per cent by selection within the existing varieties using methods from Section 8.2. Beware, however, that selection generally reduces genetic diversity.

When there are many limitations it may be worthwhile to start a Participatory Variety Selection, i.e. identify materials from outside that could be tested by the community (see Chapter 12). This could result in the replacing of local varieties or in enriching them through including selected new materials in these landraces.

When, on the other hand, a solution is not to be found with the local and new materials, Participatory Plant Breeding (PPB) will have to be the follow-up. In this case, varieties are selected that, when crossed with the local materials, are expected to yield materials that combine both the positive characteristics of the original local variety, and characteristics that solve the identified weaknesses. The distribution of responsibilities may differ according to the interest, knowledge and facilities of farmers and breeders. The latter may be called upon to act, guide or monitor the different stages. In the case of self-fertilizing crops, the breeder may:

○ cross the parent materials on-station and perform the initial screening there;
○ cross the parent materials on-station, but selection of the segregating materials is done by the farmers from the very start; or
○ assist farmers in crossing and selecting the materials.

The choice depends on a number of factors, mainly dealing with the skills and the commitment of the participating farmers, and on the ability of the breeder to mimic farmers' conditions on station.

The selection for easily identified traits such as grain colour can very well be done by farmers. In dealing with traits that are not very visible, such as disease resistance or yield, it must be borne in mind that especially the selection in early generations after the cross can be very difficult. Segregating populations is difficult to handle, especially where so-called 'hybrid vigour' may allow the heterozygous plant to perform better than expected. Selection for these plants presents no genetic gain whatsoever. In such a case it may be worthwhile for the breeder to follow a so-called single seed descent method, whereby the population is advanced without selection, until relatively homogeneous genotypes have developed, from which a useful selection (preferably in the target area), can be made.

Participatory Plant Breeding requires much more of the farmers than Participatory Variety Selection, and in particular it requires a commitment over many years.

8.3.3 Introducing participation in an existing programme
Participation can also be introduced in on-going research programmes. Where the previous section dealt with a situation where community action sought the assistance of a trained breeder, the following is commonly started from the other end, i.e. the breeder feels the need for closer co-operation.

There are different possible starting points. The most common one is the expansion of co-operation based on on-farm (breeder-managed) variety trials. Such variety trials have been included in many breeding programmes during the past 10 years in order to obtain more statistical data on new varieties. Such systems may be transformed step-by-step into farmer-managed and farmer-designed trials, with the influence of the farmer in the evaluation increasing. The result is that breeders get to know the farmers' criteria better, which allows them to select better materials for local trials.

An alternative is that farmer seed specialists are invited to the research station or a large-scale district variety trial in order to select interesting materials which they can test in their own fields. The options for follow-up could include:

○ farmers simply take what they like;
○ farmers take what they like and report their experiences on-farm back to the breeder;
○ farmers take what they like and the breeder assists in the design of the on-farm trials; or finally
○ the researcher goes to the area and carries out a detailed needs assessment based on what the farmers say they like about the selected planting material. This then becomes the start of the sequence of the breeder looking (internationally) for genetic material that may fit the local situation better, or fill the gaps.

Both these starting-points allow both partners to gauge the capabilities of farmers to handle larger or more complicated trials over time. At the same time the partners can gauge the capabilities of the breeders in handling the diversity of objectives that different farmers may have. This type of co-operation may thus develop from simple screening of a limited number of varieties to comprehensive co-operation in plant breeding.

A very important by-product of the co-operation between farmers and breeders at any level is that the latter are continually confronted with the farmers' priorities. This allows them to set breeding priorities more accurately and to work on how the multitude of farmers' criteria can best be balanced. Alternatively, the participating farmers are required to analyse the priorities and criteria that they may have used unintentionally.

PPB and PVS require a combination of technical and socio-economic skills from the breeders and a will to commit resources to this joint activity. Only with such co-operation is it possible to build successfully on the knowledge of farmers and use their skills in selecting varieties to meet their diverse needs.

9. Technical aspects of seed enterprise development

The development of seed businesses and especially the bottom-up approach of supporting farmers who specialize in seed instead of bulk production requires both seed technology skills and business economics. The principal question of demand assessment has major technical implications. This chapter introduces very basic differences between crops based on fertilization mode, storability, multiplication factor, and seed transmission of diseases as major success factors for seed business development. It also emphasizes the cost of specialization on seed, compared with producing seed as a by-product of food production.

9.1 Introduction

The commercialization of seed production is one way to improve seed supply. Sustainable commercial seed enterprises help to guarantee a regular flow of seed and should also guarantee a certain level of R&D that is directed at the needs of the customers. Such strategies are covered by the term 'seed business development'.

Seed businesses can be developed from three clearly different angles:

○ Promoting the involvement of existing (international) seed businesses in the national seed supply, either through facilitating seed imports or by allowing such companies to establish a branch or joint venture locally. Such international companies may be attracted through investment or tax incentives and by being allowed to sell their own varieties from the start.
○ Promoting the involvement of national farm-supply businesses in the marketing and production of seeds alongside fertilizers, agro-chemicals or farm tools. Such businessmen require technical assistance in order to allow them to deal with this living product, and a liberalized seed market without unfair competition from (semi-) public seed enterprises.
○ Promoting advanced farmers to specialize in seed production instead of, or next to, their traditional crop production.

Within the framework of this book, the emphasis will be on the third option. Promoting specialization in seed production may follow activities to increase the quality of farm-saved seed.

The possible negative effects of subsidizing seed
Subsidizing seed has been a common policy in many countries. The involved public seed enterprises are often not the most efficient organizations. Also,

on a smaller scale, projects should be very careful with subsidizing seed enterprises in the course of their development. Subsidies may lead to inefficiencies and insufficient attention to the needs of clients. Projects subsidizing seed can also take away the initiative of local farmers to specialize or of merchants to engage in seed-related activities.

Marketing is the basis of any commercial activity and should thus get significant attention from the very start of small-scale seed enterprise development. Seed marketing has to take several aspects into account; the most important being:

o demand for seed;
o price/quality ratio compared with other sources (especially farm-saved seed);
o supply: timeliness, distance, package size, reliability; and
o promotion.

Section 2.3.2 outlined the basic points and options for small seed business development. This chapter briefly elaborates on seed marketing in relation to seed enterprise development and the technical aspects of seed production. It intends to give some guidance only, and does not aim at being comprehensive with regard to marketing theory and economics.

9.2 The demand for seed

The development of any enterprise should start with a survey of demand and competition. The more important a crop is in a given area, and the more seed has to be used per unit area, the more the total demand for seed is (i.e. total area multiplied by the seed rate). Demand depends:

o the willingness to buy seed (instead of on-farm seed saving), which relates to widely experienced problems with seed quality and availability;
o the characteristics of other sources of seed outside the farm, such as neighbours, relatives and the grain market with respect to quality, price, and secure availability; and
o farmers' financial capacity to buy seed from a new source.

The expected demand is then calculated by including a 'seed replacement rate' (see Box 9.1). This rate is the average number of years a farmer uses and reproduces the seed before buying new seed. The replacement rate depends on the degeneration of the seed beyond acceptable limits, i.e. when yields become seriously depressed. For instance, a replacement rate of once every four or five years is quite normal for OP maize varieties. The replacement rate is normally used as a highly theoretical average which does not take into account differences among farmers.

These calculations may result in the expected 'effective demand' for seed. Further information is needed, however, before this can be translated into a good business idea. Firstly, the farmer must determine whether seed quality is a problem at all in the area, and estimate the frequency with which farmers buy seed for the given crop. Secondly, if there are problems, techniques to deal with these problems (e.g. new variety, disease-free seed, higher germination) must be available in order to develop a market. Thirdly, such problems should be overcome within the financial capabilities of the intended customers. If problems are significant, or if the solutions to such problems are very expensive, seed enterprise development is bound to fail.

  Another reason for seed demand may be the lack of reliable supply from normal on-farm sources. This generally means that there is a demand for

137

seed only after years of low yields. In a bad year for crop farming, seed production will also be poor. This is the basis of the anti-cyclical nature of seed supply and demand: when seed yields have been good and stocks are available, demand is low because on-farm stocks are also high, and vice versa. This type of demand is not a sound basis for a local seed enterprise to become sustainable.

9.3 Costs and price/quality ratio

The cost of seed is an important factor, especially when seed may be the only cash input. The chances of small seed businesses operating successfully depends both on technical capabilities to overcome existing seed quality problems and on the price of such seed. This section intends to summarize the cost of seed in relation to the size of the organization: from on-farm produced and saved seed to production by formal institutions.

Cost in relation to specialization
Seed is almost always more expensive than grain. Even grain that is bought from the market at the planting season and used for seed represents an added cost, since the grain price is likely to be higher at planting time than at harvest time, and farmers may have to sort the grain (e.g. to remove shrivelled or stained seed) before planting.

If farmers use their own farm-saved seed they have 'locked up' money in the stored seed which they could have used for income-generating activities, which represents an opportunity cost (i.e. the money they could have earned, had they used it on the next best option).

The price of on-farm produced seed of cereals or legumes may be considered the same as grain produced in the same field. This is not entirely correct, especially where the farmer has used additional time for selecting the best heads for seed, and because special care has been given to drying and storing the seed. It is difficult to express this additional cost in monetary terms, because family labour is often assumed not to have an opportunity cost (i.e. they would not have had another income had they not selected the seed). An estimation is that seed selected from a crop production field may be valued at 5 to 10 per cent higher compared to the grain produced in the same field.

As soon as special plots are assigned to seed production, which get additional attention, including fertilizers, disease control and supervision, seed production may cost up to 20 per cent more than grain from the same or similar farms. (This percentage is a common bonus for contract seed producers in formal seed production systems.)

The cost of seed increases sharply with a specialization in seed production, i.e. specialized seed farmers and specialized seed processors and distributors. The post-harvest operations may add considerable costs. Mechanized seed cleaning and seed dressing require a certain scale of operation warranting an investment in the equipment. This in turn introduces significant transport costs from seed growers to processing plant, and from plant to seed users. Also, depreciation of equipment and the use of packaging material may represent important cost factors (see Box 9.2). In practice this means an increase in cost of approximately 50–80 per cent for bulky seeds (beans, cereals) compared with grain. This means that an efficient seed production organization will have to charge farmers close to double the price of grain for centrally produced/processed seed. This cost may include the cost of internal seed quality control, but not all the cost of plant breeding, seed research and marketing. When such costs are calculated, along with interest rates and some profit, the seed/grain price ratio will increase even further. This also happens when inefficiencies have to be taken into account, for example those that often occur in public seed production organizations.

Box 9.2 Schematic presentation of cost build-up of large-scale seed production

Grain	———	100%
Farm-gate seed price	———	120%
+ transport	———	130%
+ processing	———	160%
+ processing losses (5%)	———	168%
+ packaging/treatment	———	180%

These prices do not include:

○ interest on capital
○ overheads (management mark-up)
○ marketing cost: promotion, retailer's mark-up, etc.
○ research.

Seed price in relation to multiplication factor
The cost of purchased seed may be quite high compared to on-farm alternatives when the multiplication factor is high. In some cases this may be a problem for a farmer, in other cases not at all. The problem of increased seed prices is most prominent in large-seeded self-fertilized crops, such as beans and groundnut. These crops have a low multiplication factor, i.e. the ratio between the seed harvested and the seed sown per hectare. For groundnut this ratio is very low: between 5 and 10 depending on the plant type and the cropping system (e.g. 100kg planted to obtain 600kg of clean seed after the season). This means that a farmer has to invest 10 to 20 per cent (in volume or weight) of his expected yield in seed. When the seed price is twice the grain price this means that he has to invest 20 to 40 per cent of the crop value in seed alone. This

means that output should increase considerably as a result of the use of the purchased seed. Another reason for a farmer to purchase seed may be to obtain a new variety. Particularly in self-fertilized crops (rice, beans) this is not a very sustainable type of seed demand. The profit margins on seed of those crops tend to be small. Commercially sustainable large-scale seed production with mechanized seed processing is not easy for those crops.

For crops with a high multiplication factor (e.g. sorghum at 100–150; i.e. 8 kg/ha seed sown for 800–1200 kg grain harvested), the investment in seed is relatively small (approx. 1 per cent volume). Even if purchased, with seed costs five times the grain price, a yield increase of only 4 per cent would be enough to warrant the investment. From a business development point of view, this means that commercial prospects are generally better in small-seeded crops compared to those with a low multiplication factor.

Seed price in relation to quality
Seed production may have commercial prospects when farmers face specific problems keeping their own seed. Quality deterioration may be caused by genetic, physiological, physical and sanitary causes.

Farmers may experience genetic deterioration in cross-fertilizing crops like maize or sunflower. Physiological deterioration occurs in crops like soybean where the seed easily loses viability in tropical climates due to humidity and high temperatures. Physical (analytical) seed quality may be a problem in very small-seeded crops like pasture grasses, and sanitary quality may occur when seed-transmitted diseases accumulate in a certain crop (e.g. beans, potato).

When seed producers have an effective answer to such problems, they may get a ready market at commercially interesting prices even for 'bulky' seeds. A bean seed producer may be able to supply seeds with significantly less seed-transmitted disease by growing them at a higher altitude. Soybean may be commercially interesting when seed can be produced in the off-season, thus avoiding degeneration during prolonged seed storage. The basis is that a cost-effective solution to a real problem can be offered.

General guidelines are, however, to keep production costs low. For 'bulky' seeds this means keeping processing and transport costs low through decentralized production and small-scale processing equipment.

Commercial crops
Irrespective of the multiplication factor and quality considerations mentioned above, commercial prospects for seed business development are better for market crops (sunflower, sesame) than food crops (millet, cowpea) (see Box 9.3). Cash investment in locally consumed food crops are not readily made, whereas funds to purchase seed for market crops like sunflower are more forthcoming. Secondly, quality standards for market crops often include uniformity aspects of the product (e.g. shape of tomatoes,

coloured beans). This means that seed has to be purchased when, for example, isolation cannot be guaranteed and farm-saved seed may exhibit too many off-types. Such uniformity is not often necessary in home-consumed crops, where admixtures may even be preferred. In fact this means that the seed replacement rate is usually higher in market crops.

Box 9.3 Business opportunities

- Seed enterprise development is likely to be more effective with crops with a high multiplication factor rather than with those with a low one.
- Seed enterprise development is likely to be more effective with seeds of commercial crops rather than subsistence crops.
- Seed enterprise development is particularly effective with difficult-to-produce seeds, such as biennials and hybrids.

Maize is an important crop, which has a dual purpose in many countries: both a food and a market crop. For such crops, a seed/grain price ratio of up to 4 is common; for hybrids it may go up to 16 or more. This means that more attention can be given to grading, treating, and packaging the seed (requiring high investments, and thus concentration of operations) and that a good profit margin may still be obtained.

Seed cost in relation to the ease of saving seed
Until now, we have mentioned crops of which the harvested (and con-sumed) product is basically seed, such as cereals, legumes and oil crops. For other crops, in particular vegetables, the competition with farm-saved seed may be very different.

Crops for which particular treatments are necessary to produce seed may benefit from the specialization of seed producers. Biennial crops like onion and cabbage need special skills in selecting and in growing the crop beyond the harvestable product stage. For such crops individual farmers or particular farming communities may develop the special skills necessary for producing good seeds. Such farmers or communities become known as valuable sources of seed. Commercialization of such local seed specialists may not be very hard. It must be borne in mind, however, that such minor crops require a rather large distribution area in order to become truly commercially interest-ing for such seed-producing communities. If serious plant breeding has to be undertaken in such crops, even a country may be too small a market for a seed company. The economy of scale is very important in this context.

9.4 Competition

The previous section concentrates on competition for a seed business from farmers who can retain seed from their fields. When this competition can be avoided, prospects change dramatically. One way is to introduce hybrids. Hybrids do not breed true to type and the loss of hybrid vigour may force farmers to purchase seeds every season. Because the farmer depends on the seed producer for every sowing, this allows for more liberal price setting by the seed producer, securing better profit margins. In some hybrids (e.g. maize) this loss in yield potential may not be so severe, and farmer-retention of hybrid seed does occur quite successfully in some countries. Obtaining the parents and introducing the level of production planning and quality control in order to produce hybrid seed is usually beyond the scope of projects supporting small-seed business development.

Even if a project can get hold of parent lines of hybrids, it should think hard before promoting hybrids simply because it provides commercial opportunities to develop seed production. Farmers using hybrids become dependent on the seed supplier, a situation which does not always meet with rural development objectives.

Another way to secure a ready seed market is chemical seed treatment against diseases or pests. Often such chemicals are not readily available to common farmers. When such treatment solves a particularly serious problem, there can be a ready market for seed. It must be clear that environmental protection and human health objectives have to be met in developing such business strategies.

Apart from competition with farm-saved seed, a seed enterprise has to deal with competing enterprises. This could be other small-scale enterprises, large national seed corporations or even multinational seed companies. For every competitor, an entrepreneur has to develop a strategy to develop and protect the market share. Against other local enterprises he or she may concentrate on specific quality characteristics that are important in this market, e.g. the importance of smut in sorghum. Furthermore, a lot of attention may be given to establishing a quality label, i.e. to connect the name with 'quality' and 'reliability'. Finally, the entrepreneur may develop contracts with e.g. local co-operatives to supply certain quantities every year, which will give a certain guaranteed market and a good commercial basis to the enterprise. The entrepreneur may be an individual, a small group of individuals within a village, or a co-operative at any scale.

Against national seed corporations, competition is sometimes hard, because such semi-public organizations often sell seed with open or hidden subsidies. The local enterprise may develop a policy to stress adaptation of the variety to the regional conditions, timeliness of supply and the combination of seeds and advice directed at local conditions. Also, barter arrangements may be effective to gain a market share. Against multinational seed companies a local seed enterprise may compete with a lower price and by supplying a complete package of all seeds that the farmers in a given area need, in addition to a combination of locally adapted varieties, advice and barter arrangements.

9.5 Distribution

In building up a stable market for seeds, the supply side has to meet the requirements of the demand. The quality of the seed has to match the price. Furthermore, the client (farmer) has important expectations that the supplier has to meet. The distribution system has to serve the client optimally.

A major aspect is the timing of seed supply. Farmers often like to buy seed just before planting, i.e. in many cases when the first rains fall. Seed that is available before that period must be stored at distribution points with sufficient care (low temperature and moisture, and protected from rodents), otherwise seed available after that period will not sell at all. The planning of the distribution of seed to selling points has to be done very efficiently. Problems with timeliness of supply are a major factor in the failure of public seed enterprises.

Secondly, the distance to selling points is a major success factor for selling seeds. Farmers should be able to reach the nearest selling point comfortably, having minimal transport problems carrying the seed to their farms. This means in practice that the distribution point should be near enough to be reached on foot or by bicycle within a few hours or that taxis are available to transport the seed near enough to the farm. It may be clear that where the distance between selling points and farms is short, the distribution system becomes expensive, and the influence of the supplier on, for example, storage conditions becomes smaller.

Furthermore, farmers should be able to buy the quantity of seed that they need. Farmers planting 0.5ha of maize (10–15kg of seed) should not find seed packed in 100kg gunny bags. A seed producer should pack the seeds in correct package sizes, because when the retailer opens the packs and scoops out the amount needed by a customer, all kinds of deceptions may occur. For example, a trader may mix seed with grain to make an extra profit, frustrating the farmer's confidence in the quality of the seed, which will spoil the producer's brand name. Smaller seed packets, however, increase the price of the seed.

9.6 Promotion

For any business to be successful it is vital that the consumers know the product. This is no different with seeds. The farmers have to know the supplier (brand name) and have to associate that name with the qualities of the product. This could be seed quality *per se*, or the reliability of the supply. Promotion is an essential element in seed enterprise development. The messages and the channels depend on a number of factors, for example:

○ the geographical range that the enterprise wants to operate;
○ whether promotion is aimed at existing or prospective (new) customers;
○ the type of farmers the enterprise wants to reach;

o the message, i.e. what associations does the enterprise want to make; and
o the budget the enterprise has for promotion.

There is a wide range of options available, and methods may be copied from other products. Printing colourful posters for farm-supply shops and local markets may be effective. Such posters with little text and clear pictograms may be very effective to reach uneducated farmers. Trade name, company logo and catchy slogans are good ways to develop a market. Trademarks are very important in word-of-mouth promotion.

●●●

Seed doesn't cost, it pays

●●●

Leaflets which combine trademark-promotion and product information may be effective for literate farmers. Promotion during local agricultural fairs, for example, on world food day, can attract a lot of attention. Radio or television messages, preferably in the form of documentaries, may reach yet other types of farmers, e.g. the town-based landowners.

●●●

Better crops with seed from Hobbs

●●●

Seed promotion also takes the form of demonstrations and field days. This is particularly effective to introduce new varieties. This could also be done by distributing small packets of seed, for example together with the purchase of chemical fertilizer. It is extremely important to direct promotional activities to the specific target group that the product is intended for. Do not, however, forget the easiest promotional tool: the packaging material. It should be attractive, informative and with a clear brand name and logo.

●●●

Good seed, indeed, is what you need!

●●●

9.7 Seed production planning

When an estimate of the demand is made, seed producers can plan their production. Production cannot be raised in a very short time. The bottlenecks are:

o the availability of sufficient land, with farmers who are able to produce seed of sufficient quality;

o the availability of sufficient quantities of high-quality seed to plant for the production of the marketable seed;
o sufficient funds to produce the seed; in practice to finance the production (or purchase from contract growers) until revenues from seed sales are received; and
o in some cases the seed processing and storage may be a bottleneck.

When seed businesses are very small and the main seed crops are cereals and legumes, the first aspect may not be a major problem. Seed is merely grain produced with great care and experience. Seed which is not sold is consumed. When demand is higher, the owner of the business will know of good farmers in the area whose grain can also be sold as seed.

This section concentrates on the second aspect and is based on the notion from formal seed systems that seed produced is always of lower genetic quality than the seed sown to produce it. This has resulted in the generation system: small quantities of very tightly monitored 'breeder's seed' (pre-basic seed) is multiplied through a limited number of generations to large quantities of certified seed. This means that the quantity of certified seed to be produced has to be planned several seasons ahead. For smaller-scale seed production enterprises also, planning is crucial. Seed as a living matter cannot be stored indefinitely. Insufficient stocks spoil the name of the seed producer with regard to the reliability of supply; over-production will be a cost factor to the enterprise which will have to be recovered by raising the seed price. Also, small-scale seed producers will face degeneration of the variety of the seed quality when the seed is just bulked every season without careful selection, i.e. some form of generation system should be applied when the amounts produced increase.

The multiplication factor is the main tool for calculation in such a planning process. The multiplication factor is the amount of seed harvested divided by the amount of seed planted. Seeds with a low multiplication factor will need a more careful planning of generations, compared to those with a high multiplication. Table 9.1 presents the planning of the production of enough seed for 100ha of sorghum and bean crop production in the year 2005. It is assumed that average bean and sorghum yields are

Table 9.1: Production planning through various generations of seed for 100ha of bean and sorghum crop production in the year 2005

	2004	2003	2002	2001
Beans	10ha planted 10 000kg	1ha planted 1000kg	0.1ha planted 100kg	0.01ha planted 10kg
Sorghum	1ha planted 1000kg	0.01ha planted 10kg		

145

approximately 1t/ha. Sowing rates are 100kg/ha and 10kg/ha respectively. This means that the multiplication rate of bean is 1000/100=10, that of sorghum is 1000/10=100. For 100ha of bean production in 2005, 100ha × 100kg/ha=10 000kg seed needs to be harvested in 2004. With a multiplication factor of 10, this means that in 2003 1000kg seed should be harvested to be planted in the 2004 season, 100kg in 2002 and 10kg in 2001. These amounts can be linked to the planning of the land needed for planting the seed crop. The multiplication factor is also reflected in the land needed for the seed crop. Planting 1000kg of seed in 2004 requires (with the given sowing rate of 100kg/ha) 10ha of land. In 2003 1ha (10 000 m^2) is needed, in 2002 and 2001 respectively 0.1ha (1000 m^2) and 0.01ha (100m^2, or a field of 10 by 10 m). For sorghum a similar calculation can be made.

This means that bean seed production and the choice of the initial seed stock should be planned many years ahead. As early as 2001 the demand the enterprise will face in 2005 must be anticipated. This is not very practical for a small-scale seed business, which would be better advised to start a strict build-up of a 'clean' seed stock for example every five years. This is done by selecting from the bulk production a small seed sample of around 50–100 plants, which is planted separately the following year. The crop grown from this sample should be carefully checked for off-types and diseased plants. Very strict row selection should be practised in the first year (see Section 8.5) and negative mass selection in the following years. Such a routine build-up of the seed stock ensures reasonable quality seed all the time. In the mean time, the bulk is multiplied and available for distribution.

In a similar way, a seed producer can calculate the time needed for enough seed of a new variety to be available for marketing. When a small amount is received from a research station, for example, it may not be wise to start promotional activities directed to that new product straight away, because it may take a few years before it can be put on the market in reasonable quantities. Promotion would lead to demand which, when supplies are insufficient, will lead to a reduced confidence in the enterprise.

This pre-supposes that the demand can be estimated in advance quite precisely. This is often not the case, especially where demand depends on how much seed has been saved on-farm. In this case, demand is suddenly very high after a poor growing season. In order to cover such eventualities, seed producers may have to produce more than is necessary in average seasons each year, which is very costly. The register of trustworthy farmers who have bought the seed, and whose crop could be turned into seed in case of shortage has to be kept in mind.

10. Seed quality testing

An accurate analysis of seed quality parameters provides important information on the limiting factors in seed production, such as disease control, drying, storage, etc. Regular testing of seeds in a standardized manner will provide the required data. This chapter presents such standardized seed testing methods and suggests how to use them in farmers' seed production support programmes.

10.1 Introduction

A proper analysis of the quality of both farm-saved and locally distributed seed, and seed produced by the formal sector, provides an insight into the limitations of the existing seed systems. This in turn offers the challenges for improvement that this book intends to present tools for.

Some assessment of seed quality can be ascertained through surveys that record the perception of farmers regarding the seed that is available to them. At some stage, however, there is a need to measure seed quality quantitatively. Measuring seed quality is useful to validate the farmers' perceptions, and to monitor the effects of any attempt at improvement (see Section 12.5). Furthermore, measuring seed quality is an important tool in adaptive research, and in designing national seed policies and business planning for established and new seed enterprises.

Finally, measuring seed quality is an important tool for formal seed suppliers to analyse weaknesses in the chain from breeder's seed production to seed selling. This monitoring of seed quality stimulates the giving of the seed extra care, and provides reliable data for appropriate and timely management decisions (see also Section 5.3).

Where seed is produced with local facilities, seed testing is not always considered the first priority since the specialized equipment is not available and some limitations may seem obvious anyway. Sound data can, however, help to set priorities, assess progress, and draw earlier attention to seed deterioration. Rather like integrated pest management strategies that depend on the careful monitoring of insect populations and disease development in the field, this enables the right decisions to be taken in time.

ISTA, the International Seed Testing Association, has developed standardized methods for seed testing which allow comparison of results of seed tests made in different places and times (ISTA, 1999). Ideally testing is done in internationally accredited seed laboratories which are established in many countries. Alternatively, testing can be organized locally at different levels of precision. This section attempts to describe the most important standardized tests and their adaptation to local conditions. The most important tests measure:

○ moisture content;
○ purity;
○ germination and vigour;
○ health; and
○ genetic composition.

For all these tests the sampling procedure is an often overlooked, but vital component. Finally, in some cases 'post-control' can be given some consideration.

10.2 Sampling

The problem
The value of test results depends to a large extent on the degree to which the seeds tested represent the total seed lot. Proper sampling may prevent inconsistencies and inexplicable results. For example, small seeds tend to collect at the bottom of bags, and germination may be lower in seed bags that touched a damp wall during storage. When sampling seeds from the top of a bag only, results will not therefore be representative.

The goal
Random samples are taken from a seed lot and combined to obtain a sample that is representative for the entire seed lot. The samples are mixed and from this composite sample the desired quantity for sub-samples can be taken.

However, sampling techniques are different for the different tests we may wish to carry out. Therefore, a distinction has to be made for situations where a representative sample is taken from a lot which can be used for further testing; and samples that are needed for moisture testing; for a number count test; and for a germination test or grow-out test.

The sampling procedure
Normally the seed will be in jute or woven polythene bags when ready for sampling. In the absence of simple sampling equipment such as sampling triers (spears), a few bags (at least five) are randomly selected and some handfuls of seed are taken from different locations in each one. If it is too difficult to reach the bottom by hand, such as with rice seed, the sampler should remove some of the contents of some of the bags to be able to reach the bottom half. Bags may also be tipped out into another bag if one suspects that the lower half may contain more debris, e.g. stones. Thorough

mixing of these handfuls produces the composite representative sample and the desired quantity can be taken from it. Detailed sampling instructions for the specific tests are given in the following paragraphs.

10.3 Moisture content

The problem
Too high moisture content seriously threatens the quality of seed. Measuring seed moisture content is a relatively easy and effective method to assess the potential to store the seed. Safe moisture contents in relation to temperature for a number of species are given in Table 10.1. Rapid moisture determination methods are needed when decisions have to be taken in the

Table 10.1: Safe storage conditions for the major crop groups (generalized within species)

Crop group	Safe moisture content (%)	Safe relative humidity* (%)	Maximum storage temperature (°C)	Drying temperature**and time + grinding requirements for moisture tests
Cereals	12	30	40	130°C/2h/grinding
Maize	12	30	40	130°C/4h/grinding
Chickpeas/lentils	13	30	40	130°C/1h/grinding
Beans	8	30	25	130°C/1h/grinding
Brassicas	8	30	30	103°C/17h/no grinding
Oil seeds	8	30	25	103°C/17h/no grinding
Sunflower	8	30	25	103°C/17h/no grinding
Vegetable seeds	8	30	25	103°C for 17h or 130°C for 1h/no grinding
Forage grasses	12	30	40	130°C/1h/no grinding

* Relative humidity (RH) of a room can be determined with the wet bulb/dry bulb thermo-hygrometer: a twin thermometer of which one bulb is covered with a wet cloth. The temperature difference between the two thermometers is a measure of the RH of the ambient air, and can be converted into RH with the table which goes with the thermometer set.
** The drying temperature depends on whether the seed contains oil or not.
Note: these temperatures are recommended for seed moisture testing, they are far too high for drying seed!

149

field or when seed lots arrive from the field. If too moist, the seeds must be dried immediately. Monitoring of the drying procedure and of seed in storage with a portable moisture meter is recommended. A meter will certainly be too expensive for individual farmers, but may be collectively acquired, e.g. through a project.

The goal
To determine the moisture content of seeds to establish their potential for storage and the need for (re)drying. This is done either by (1) measuring the absolute moisture content directly by heating the seed and measuring the weight loss; or (2) indirectly measuring the relative humidity of the air in equilibrium with the seeds; or by (3) measuring the conductivity of (compressed) seeds.

Sampling
If sampling for moisture testing, a quick procedure must be followed which prevents the seeds from drying during the procedure. This means that when a bag is opened for sampling as indicated above, a small extra handful of material has to be taken and put into a plastic bag or tin. Be sure to use intact plastic bags of considerable thickness. This procedure is repeated with a number of bags. The sample, which will preferably contain at least 100g of seed, is sealed well, with a label inside the bag as well as outside, and tested as soon as possible. The seeds from this bag cannot be used for other types of testing, because of the risk of the seeds having deteriorated.

Rapid tests
Easy and rapid methods are provided by portable rapid moisture testers. These battery-operated meters give results almost instantaneously. The apparatus reads the conductivity of the air surrounding the seed. In a balanced situation there is a fixed relationship between humidity in the air and the seed, and the reading on the apparatus can then be translated into a moisture content by using a conversion chart for that species (supplied by the manufacturer) (If no conversion chart is available, see Box 10.1). More expensive rapid moisture testers for laboratory use and with higher accuracy than the portable ones are commercially available.

Oven test
The method commonly used by official seed-testing stations involves a small amount of seed being weighed accurately and then heated to a prescribed temperature. The seed is weighed again after a period, and weight loss is used to calculate the original moisture content. The danger of overheating the sample exists: if too hot, other volatile components also may evaporate, such as etheric oils. This results in over-estimation of the moisture content. Temperature should never exceed 103°C for oil-containing

seeds (Table 10.1) or 130°C for starch-containing seeds. Checking with a thermometer is therefore necessary when using the apparatus for the first time for a given species.

Larger seeds (those the size of sorghum, wheat grains and larger) have to be coarsely ground, cut, or broken up. The quantity tested is 10g, usually in two replicates of 5g. Coffee grinders cannot be used because they heat the seed, inducing heavy evaporation. Household ovens cannot be used either, because of their very high temperatures, even at the lowest setting.

It should also be noted that for wet seeds (over 20 per cent moisture content) all moisture tests tend to become less reliable, and a special pre-drying procedure should be followed.

10.4 Seed purity

The problem
Weeds and diseases may endanger the crop in successive generations. Therefore a distinction is often made between weeds and diseases which are merely a nuisance and those which pose a real danger, the so-called noxious ones. The purity test as described here concentrates on weeds and

other foreign seeds. Purity analysis is rarely done in local seed initiatives. The amount of inert matter (chaff) is, however, important when seed production is commercialized (customers do not want to pay for chaff) and especially when seed is infested with a lot of weed seeds.

In official testing, one distinguishes 'purity tests' proper from 'number count tests'. The purity test is to determine exactly the composition of the seed lot, including weed seeds, chaff, empty seeds, debris, etc. The number count test is to count the number of seeds of other species present in the lot. For local seed production, the purity test is not very relevant and number count tests are described instead.

Box 10.2 Weeds: a nuisance or a pest

If present in sowing seed, some weeds may endanger a good harvest, but not all weeds are this bad. One usually distinguishes between weeds which are merely a nuisance, and those which pose a real danger, the so-called noxious weeds.

Noxious weeds are those which multiply more rapidly than the crop itself or are very difficult to remove from the harvested seed or from the field. Examples are seeds of *Rottboelia exaltata* in rice, which is difficult to remove from the seed, *Apera spica-venti or Sporobolus* species which have a multiplication rate of 100 in a cereal crop, and *Centaurea* species which may be very persistent through underground suckers. Other examples are parasitic weeds such as *Cuscuta, Orobanche* and *Striga.* Finally, there are the poisonous seeds of, for instance, *Ipomoea, Agrostemma* or the dangerously spiny seeds of *Emex spinosa.*

Particularly noteworthy examples are the fungal sclerotia of *Claviceps purpurea* (ergot) in cereals and *Sclerotinia sclerotiorum* in sunflower and cabbage. The first poses a serious health risk if present in flour, the second is a special form of seed-borne disease (see 12.6 below).

The goal
To count the number of foreign seeds (weed seeds and seeds of other crops) in a given sample in order to estimate the numbers sown per unit area.

The number count test
Sub-sampling. For a number-count test some 25 000 seeds are usually tested. Because counting this number of seeds is impractical, a sub-sample is prepared of a weight which is known to contain about this number of seeds (Table 10.2), with a maximum of 1000g. By experience one learns how much is needed for the most common species. The required amount is obtained from the seed sample bag by pouring the entire contents of the bag on to a large sheet of paper, e.g. a newspaper, or a blanket. Then

Table 10.2: Sample weights for number count tests

Crop	Weight to be tested (g)
Pearl millet	150
Rice	400
Lentil	600
Sorghum	900
Wheat, barley, oats	1000
Chickpea, grams	1000

spoonfuls of seed are scooped randomly from the mass, until the desired quantity is obtained.

An elegant alternative is the random cups method, whereby a set of small empty containers (such as small tins) is placed randomly on the cloth. The seed is then poured evenly over the entire surface. Afterwards a random selection of cups is taken to produce enough seed for the working sample.

The number count test is particularly relevant in crops that are machine-harvested. This applies to most field crops, and is less useful for most vegetable crops, especially if the seeds are taken from hand-picked fleshy fruits, such as tomatoes, egg plants or cucumbers. Also, large-sized seeds like beans and maize usually do not contain other seeds.

Analysis and calculation.
○ Pour the entire working sample on to a flat surface, using rulers or other flat rods in order not to spill the seeds, or on to a large flat tray especially constructed for the purpose (a piece of very flat board, such as plywood, of about 40 × 30cm, with edges from laths of 2 × 2cm).
○ Put the seeds on one side of the table, then move them all, seed by seed or in very small quantities (maximum of about 5–10 seeds) from one side to the other.
○ Any seed which is not of the desired type should be put to one side in one of several containers that you have ready (e.g. the tins used for sampling).
○ Put the rejected material in the containers, for example all weed seeds together in one mix, seeds of other crops in a second tin, off-type seeds which belong to the same crop but which are different and not desirable in a third tin, and finally fungal bodies and insects in a fourth. If desired, one can add categories which need particular attention, such as seeds with a clear insect hole, discoloured seeds affected by a certain disease, etc.

○ Count the number of seeds of the various species. For this purpose, it is necessary to recognize these seeds. In the absence of a seed collection, this knowledge can be obtained only by careful observation during the growing season (see Box 10.3).

○ The number of seeds of each species is noted in a seed-testing notebook, next to the total sample weight, the date, and the name of the analyst.

Box 10.3 Making your own seed collection

The capacity to recognize plants from their seeds, seed identification, is a specialization that can be developed over time, and which should be supported by a seed collection.

To this end, collect seeds from all plants you encounter in the field, put them in paper bags while you work, and write the name of the plant on it, or if you do not know the plant, take one with you to show a knowledgeable farmer or a weed specialist. The plants should best be kept in a herbarium (bundle of folded newspapers), dried well and kept in a dry box. Note vernacular names and other details on a label which you attach to the plant.

These dried plants can also be used to show to colleagues and farmers during field schools. Taking good photographs of weeds is difficult, and plants may serve the purpose even better. Even without knowing the official name of the weed, it can then still be recognized.

10.5 Germination capacity

The problem
The main purpose of sowing seed is to produce a healthy and uniform crop. Germination capacity and vigour are two quality criteria which strongly determine field establishment. Germination capacity is defined as the percentage of normal seedlings produced by pure seeds.

Vigour can be defined as the capability of a seed lot to perform comparatively better under unfavourable conditions than seeds of another otherwise totally similar seed lot. Although this second criterion seems to be most relevant for farm conditions, it has proved very difficult to devise appropriate tests with consistent test results (see Section 10.6).

The goal
To determine the intrinsic seed germination capacity of a lot, the percentage of normal seedlings which can be produced by the lot under more or less ideal conditions.

The germination test
Sub-sampling for germination tests and field trials. For germination tests usually about 400 seeds are used. They can best be obtained by mixing the

154

seed (obtained by sampling, as described in Section 10.2) on a table, and dividing the heap into as many 'sub-heaps' as one needs replicates – normally four – and counting 100 seeds from each heap at random.

It is generally accepted that only normal seedlings are of agricultural significance, and not just any germinated seed. The seedling should not show important defects and should be able to become a healthy competitive plant. The general rule is that all plants should be of about equal size and development at evaluation, that all essential structures must be present, and that no diseases are observable. Root and shoot development must be in balance.

Suitable media for growing seedlings. Petri-dishes, with a layer of filter paper and kept at ambient conditions, are often used for informal seed testing. This method, however, is not advised: the atmosphere is too humid and sometimes too hot, and is totally unsuitable for seeds larger than e.g. cabbage seed. Fungal infections often spread over the whole dish, while maybe only one seed was infected.

There are basically two formally accepted media on which seedlings can be grown: filter paper and sand. Filter paper is too expensive because it cannot be re-used, and other paper like tissue or newspaper is unsuitable (soggy and poisonous to the seed). Sand can be used if it is clean, e.g. sand from a river bed. Soil from the test site is not usually suitable. A relatively cheap alternative is provided by small guest towels, preferably white, which can be used for the so-called roll-towel method. These towels have to be washed and sterilized after use, but once bought they can be used for a number of years. Seeds are sown in replicates of 50 to 100 seeds, usually 400 seeds in total.

Test conditions. The easiest method is to put the seeds in the shade under a roof. Never carry out the tests in direct sunlight, because they will dry out in a matter of minutes.

Each species has its optimum temperature for germination (see Table 10.3). If results are to be compared between seasons and years it is

Table 10.3: Germination temperatures for some major crops

Crop	Optimum temperature range for germination (°C)
Maize, rice, sorghum	20–30
Pearl millet	20–35
wheat, barley, oats	20
Chickpea	20–30
Lentil	20
Soybean, beans, grams	20–30

important to standardize test conditions as much as possible. Ideally incubators are used, which can be set at the right temperature. Some manufacturers modify household refrigerators or wine coolers to meet the requirements. If this apparatus is not available, an air conditioner may stabilize the temperature in a room, but tests will dry out quickly. Creating a moist environment by making a plastic tent around the tests can be effective.

Analysis and calculation. Seedlings are evaluated when the majority of the seedlings have reached a stage where all essential seedling structures are visible, which usually means that the cotyledons have expanded and the terminal buds (dicots) are visible or when the first true leaf has perforated the coleoptile (the sheath of monocots). The number of normal seedlings is counted and the percentage calculated. If a large number of seedlings seem slow, evaluation may be delayed. If a large number of seeds do not germinate they are either empty or dead seeds (break open and see if empty or rotten) or the seeds are dormant (seed interior seems perfectly normal). In the latter case special measures can be taken to break dormancy.

10.6 Seed vigour

The goal
To discriminate between good and bad lots and to get an impression of field emergence under local conditions. This is done by germinating a sample of 200–400 seeds to determine the percentage of successfully emerged seedlings which can be produced by the lot under less favourable conditions.

The seed vigour tests
It would appear to be the easiest thing to sow seeds in soil and under conditions comparable to those encountered in the field. Usually the climatic conditions when one wants to test the seeds differ considerably from those when the crop is actually sown (dry season and rainy season respectively). The best approach is to try to define the conditions that are prevailing and include factors that are limiting field emergence, such as drought, excess water, soil structure. Once properly defined, one can manipulate conditions in such a way that enough stress is given to discriminate between good and bad lots.

A simple measure to improve the tests would be to put the soil in boxes so that the tests can be moved out of the way of occasional excessive rains which would otherwise ruin the whole experiment. A transparent roof over the tests can serve the same purpose. One must make sure that rodents and birds have no access.

Particular attention must be given to the soil structure: if watered after sowing, the soil may become too wet and the gas exchange of the seeds will be disturbed. It is advisable to add ground brick or other intert material to the soil to guarantee aeration, and to provide the seeds with a good deal of moisture from the start.

Because of all these difficulties, it is recommended to plant at least four replicates of 100 seeds each.

10.7 Seed health

The problem
Seed crops may be affected by many diseases, but not all of them are seed-borne. There is a number of soil-borne and air-borne diseases which are a problem to the plant but do not necessarily affect seed quality. Examples are *Pythium* ('damping off'), which is soil-borne, and *Phytophthora* (late blight), *Perenospora* and *Oidium* (mildews) which are air-borne.

The goal
To determine the presence of seed-transmitted diseases on or in the seeds. To do this, many different tests exist, depending on the pathogen and the host. The tests for viral and bacterial diseases cannot be performed without expensive equipment, but for fungal diseases, which are the most prominent diseases in most field crops, the freeze-blotter test can be performed with relatively small investment. The test is based on killing the seed by deep-freezing after inhibition, development of the fungus, and determination of the kind of fungus with a binocular microscope.

Freeze-blotter tests
Seeds are placed on wet blotting paper, and allowed to absorb a maximum amount of water (12h). They are then placed in a deep-freezer for a period of 24h, and subsequently moved to a dark moist environment which prevents the seeds and the fungus from drying out. After some days fungi will have developed. Because the hyphae of most fungi look the same, the fungus must bear spore capsules for a proper identification. Sporulation can be promoted with UV light (black light). In theory, tests may also be placed in light, avoiding direct sunlight, e.g. in plastic petri-dishes. Glass should not be used, however, since it effectively filters out the UV light; for the same reason they should not be behind a glass window! Such test conditions in the natural environment are hard to control and drying out of the medium is a serious danger.

Identification of the fungus needs special expertise. The trained analyst will still need several reference materials and books for a reliable test result. Pathogens (and saprophytes) which can be detected by this method include: *Alternaria, Ascochyta, Botrytis, Cercospora, Cochliobolus*

(Bipolaris), *Didymella*, *Fusarium*, *Leptospaeria (Septoria)*, *Macrophomina*, *Monographella*, *Phoma*, *Phomopsis*, *Pleospora*, *Pyrenophora*, *Pyricularia*.

It must be stressed here that if a disease has been found it does not necessarily mean the disease will show up in the field. There are many buffering and antagonistic effects in the soil which cannot be predicted by the test.

10.8 Genetic composition

The problem
Genetic seed quality has a dual meaning: the value of the variety for a particular use, and the conformity of a seed lot to the variety. The first meaning is dealt with in the sections that introduce variety maintenance and crop improvement: (participatory) variety selection and plant breeding. Seed testing can contribute to the identification of varieties and off-type counts, but it should be clear that such tests are particularly important in formal seed production systems (see also Section 5.2.4).

The goal
To establish varietal identity and uniformity of a seed lot. This is done by a thorough description of seed characteristics which allows trained seed analysts to identify whether a submitted seed sample conforms with the variety as labelled, through comparison of the seed with a detailed morphological description or a reference sample, or by using specialized methods (see below).

Finally, a grow-out test (the so-called post-control) can confirm the results or add further varietal characteristics to the test.

Standard laboratory tests
The purity sample that is used to establish the quantities of other crop seeds, weed seeds and inert matter can also be used to establish the variety and off-type counts. This can be done only when seed characteristics, such as colour and shape in beans, clearly differ. Smaller characteristics, such as hilum colour in pulses, and the shape of the rachilla (stalk of grass florets) can be used to identify a variety. Such methods may allow a seed analyst to identify a number of varieties. In Sri Lanka, all the important rice varieties can be identified on seed characteristics.

Secondly, the germination test may be used to identify additional varietal characteristics, such as coleoptile colour, shape and the hairiness of leaf blades, etc.

Additional laboratory tests
A very easy tool is the near ultraviolet 'black light' lamp, which may show differences between varieties in the fluorescence caused by this light by

different parts of the seed or seedling. This method is used, for example, in peas, *Vicia* beans, oats and *Lolium* grasses, but has limited applicability.

Chemicals, such as phenol or sodium hydroxide may give differential stains in seeds of different varieties (e.g. in hard-to-identify cereals and grasses).

Finally, biochemical (electrophoresis) or even molecular tests are done to identify varieties and establish off-type counts.

Field tests

For field tests only a few plants are required to establish conformity to the variety, but a large numbers of plants may be needed to count the number of off-types. For the latter, a minimum of 500 plants per sample are used, depending on the required (statistical) precision.

Table 10.4 refers to sources where more information on seed tests can be gathered. Also, various handbooks and worksheets on seed testing have been published by both ISTA and AOSA.

Table 10.4: Details about the seed tests

	AOSA	*ISTA*	*Burg et al.*
Sampling	1	2/2a	8.1
Moisture	–	9	7
Purity	2	3	8
Number-count	3	4/8	8
Germination	4	5	9

Further information about the tests described in this chapter can be found in the above sections of AOSA and ISTA Rules, and in van der Burg, W.J. et al., 1983

PART C

WORKING WITH FARMERS

11. Problem diagnosis

11.1 Introduction

The description of local seed systems (Part A of this book) and technical information on seed production (Part B of this book) are meant to support development-oriented action. Development projects require an assessment of the present situation, and the planning and implementation for action.

'Development-oriented action' refers to all types of experimentation or events organized to support development, such as field days or fairs. This action can be broken down into a sequence of steps:

(1) orientation: involving assessment of the present situation, and identification of the constraints;
(2) planning;
(3) implementation;
(4) evaluation; and
(5) continuation.

At the end of each experiment or event there is a reassessment of existing problems, and priorities, overlapping with step (5) (continuation). The experimentation may have resolved problems, and others may have emerged as a priority. This reassessment will lead into a new cycle, repeating the steps of planning, implementation, evaluation (see Figure 11.1) In other publications the steps may be named or numbered slightly differently, but they follow in essence the same process. This chapter will elaborate on these steps.

The sequence of steps described here assumes an increased level of farmer collaboration, but is still aimed at a technician-researcher, i.e. an outsider who initiates and facilitates actions. Organizations which have their roots well integrated in farmer-communities, may give different weight to each of the steps. For example, orientation, as described here, may not be relevant or may become relevant at a later stage; instead, group discussions or other participatory tools may be preferred.

11.2 Orientation

11.2.1 General orientation
A general orientation involves discussing the farming conditions with key informants; local authorities, extension workers, farm supply merchants,

STEP
1   **GENERAL ORIENTATION**

- **formulation** of the objective: general orientation
  on use of seeds and varieties in the region
- **planning** of the interviews and study

- **collection** of the information:
  interviews key informants & secondary sources
- **systematize** and analyse information

- **conclusions**: 1) gaps in information, 2) constraints

narrowing down the scope, focusing on the objective

INFORMAL SURVEY

- **formulation** of the objective: 1) collection lacking
  information, 2) validation of assumption
- **planning** of the interviews

- **collection** of the information: interviews

- **systematize** and analyse information

- **conclusions**: 1) gaps in information, 2) constraints

narrowing down, reorientation

DEVELOPMENT – ORIENTED ACTION

STEPS
2/3/4

DIAGNOSTIC SURVEY	CASE STUDY	COLLABORATIVE EXPERIMENTATION	OTHER ACTION
• **formulation** of the objective: defining information needed • **planning** the survey	• **formulation** of the objective • **planning**	• **formulation** of the objective: reseach question • **planning** the experiment (treatments, data collection)	• **formulation** ... • **planning** ...
• **collection** of the information: interviews • **systematize** and analyse information • **conclusions**: 1) gaps in information, 2) constraints	• **collection** of the information • **systematize** and analyse • **conclusions**	• **collection** of the data • **systematize** and analyse data • **conclusions**: 1) gaps in information, 2) constraints	• **collection** ... • **systematize** ... • **conclusions** ...

STEP
5

| Diffusion of Results | Diffusion of Results | Diffusion of Results | Diffusion of Results |

STEP
1

reorientation, addressing new problems

STEPS
2/3/4

DIAGNOSTIC SURVEY	CASE STUDY	COLLABORATIVE EXPERIMENTATION	OTHER ACTION
• **formulation** ...	• **formulation** ...	• **formulation** ...	• **formulation** ...
• **planning** ...	• **planning** ...	• **planning** ...	• **planning** ...
• **collection** ...	• **collection** ...	• **collection** ...	• **collection** ...
• **systematize** ...	• **systematize** ...	• **systematize** ...	• **systematize** ...
• **conclusions** ...	• **conclusions** ...	• **conclusions** ...	• **conclusions** ...

STEP
5

| Diffusion of Results | Diffusion of Results | Diffusion of Results | Diffusion of Results |

STEP
1

reorientation, addressing new problems

CONTINUATION

Figure 11.1 Development through experimentation and action-support is a re-
petition of cycles which each consist of similar steps: orientation and problems
assessment, planning and implementation of experiments and action, evaluation
and continuation are repeated.

164

key-farmers and their family members. This interviewing is complemented by the study of secondary information sources, i.e. statistics, reports and relevant literature.

At an early stage of orientation the seed system should be considered in the context of its environment: the farming system, the farmer and the agro-ecological and socio-economic conditions (Box 11.1). It is important to know the weight of seed-related problems compared to other problems. This is also the stage at which it is most difficult to present a structure for the work or general guidelines: the researcher needs to have a good general understanding of agriculture and seed issues, and has to be creative in

---

Box 11.1 General checklist for an informal survey of the farming and seed system

FARMING SYSTEM
o Character of the average farming system: which crops are grown, how many cropping seasons, rotations, animals?
o Agro-ecological environment: climate, soils, fertilizer and water availability, variation in time and space.
o Socio-economic environment: markets, price fluctuations, government policy, household size, land tenure, ethnic groups, household size, labour peaks.
o Variation between farming systems: is there much difference between farming systems in the region?
o Changes in time: has there recently been much change in crops and varieties planted, or in other farming activities?
o Bottlenecks: what do farmers see as the most important problems in their agricultural production?

VARIETIES
o Which varieties are planted and why are they good/bad: local varieties, improved varieties?
o How many varieties does a farmer grow?
o Is there much change in the use of varieties?
o How many new varieties have been introduced?
o What happened to the old varieties – have they disappeared?
o What is the reason for the change (introduction of new varieties, new diseases, changing requirements of the farmers)?

SEEDS
o Are there problems with seed quality?
o Where do farmers get their seeds from?
o Why is the farmer using seed from other farmers or elsewhere?
o How do farmers select seeds?
o How do farmers store seeds?

---

finding the right key persons and secondary information sources. The interviewer should ask the right questions and follow-up on interesting comments when relevant. This orientation phase can be successful only when the interviewer is open-minded, has a sincere interest and a sound general agricultural knowledge.

11.2.2 Further orientation – an informal survey

After the general orientation, the target area and target group have to be defined. An informal survey can be used to confirm the picture emerging from the first orientation and to collect additional information. The objectives of the informal survey should be:

○ to verify assumptions, such as those concerning the homogeneity of the defined target area, the characteristics of the farming systems, the use of seeds and varieties and the basic information on possible constraints.
○ to fill in gaps in the information, such as an indication of how many sorghum varieties a farmer plants, whether poor and rich farmers grow the same varieties, whether landraces are managed differently from modern varieties, etc.

Formulating these assumptions and questions before the start is essential for a well-focused informal survey.

Quantitative data resulting from an informal survey, for instance on the use of varieties and seed, can be used to support a project proposal or in discussions with officials. The number of farmers to be interviewed depends on the size of the area, and differences between farmers, (large vs. small farmers, poor vs. rich farmers, male vs. female farmers). At this stage an informal survey can include a relatively small number of in-depth interviews or a larger number of interviews providing a sketch of the existing situation (no more than 50). Surveys require considerable resources, and these may be used more effectively at a later stage, when objectives are more focused.

A checklist is often useful for informal, semi-structured interviews, as a guideline for the conversation with the farmers and to keep track of the information that needs to be collected (see Box 11.1). A checklist is especially necessary when the survey is carried out by more than one team of interviewers. It is also important to elicit farmers' views rather than inferring 'expected' answers (see Box 11.2).

After the informal survey, it should be possible to give a general description of:

○ the use of varieties, seed exchange, seed sources, farmers' seed practices or variety preferences;
○ priorities and constraints for the target farmers;
○ the strengths and weaknesses of the local production and seed system;
○ remaining gaps in information.

166

Box 11.2 Semi-structured interviews and open-ended questions

At the heart of all good participatory research and development lies 'sensitive' interviewing. This involves considering one's body language, good listening and balanced questioning.

Open-ended questioning is a method for obtaining and recording farmers' spontaneous reaction. It is important to formulate the questions in a neutral, non-leading way. Six key words used in the formulating open questions are: Who, what, why, where, when, how?

'What do you think of it?'
'Why is it difficult?'
'How did you find out'

The interviewer may want to ask specific questions about certain aspects, but this is best done after the farmer has had an opportunity to comment freely.

While leaving room for the farmer to react spontaneously, it is also important to direct the flow of comments from the farmer, but this should be done in an unobtrusive way. This can be done by rephrasing or repeating the farmer's comments in the form of a question. This can invite the farmer to expand on the particular theme while it also functions as a cross-check. Examples of phrasing to stimulate and direct the farmer's comments are:

'Could you explain that to me?'
'Can you tell me more about that?'
'So, the variety resists drought?'

From: Pretty et al., 1995, *Participatory Learning & Action.*
Ashby, 1990, *Evaluating Technology with Farmers.* A handbook.

After discussion with the group of researchers-technicians, a short report should help in defining further steps. Chapter 12 deals with diagnostic surveys which address issues in greater detail.

11.3 Planning and implementing action

Formulating the objective
At the end of the orientation phase, further action can be developed, based on the findings (gaps in information, constraints). This should start with problem definition and the setting of objectives in order of priority. Problem definition is carried out jointly by the team, or together with farmers, and involves describing a constraint, for example a problem with seed germination, a general lack of good seed, or high disease pressure in the crop. The objective can be formulated in the form of a question to be answered or a goal to be achieved.

For example, if the survey indicated that at planting time the seed germination is low, the formulated question could be: how can we evaluate the effect of local storage on the seed quality? The formulated goal is similar: evaluate the effect of local storage on seed quality. Action can be discussed e.g. comparing germination of seed which was stored by the farmers and seed which was stored elsewhere or in a different manner. Once it is proved that local storage conditions are the explanation of low germination, the next cycle of action can be based on the question 'how can we improve the storage of the seeds, or on the goal 'identify possible improvements in seed storage'. Again, action can be planned and implemented, results evaluated and shared.

Planning the action
Once the objectives are defined, the actions needed to reach the answer or goal can be determined. A workplan should be prepared, including a clear time schedule. This workplan is the basis for calculating the required budget and staff requirement. Overestimating what can be achieved within the available time and budget is common. When the budget is too high, the work-plan and the objectives may have to be adjusted (see Figure 11.2).

The type of objectives and the actions depend on the results of the orientation phase and on the expertise and mandate of the organization. In situations where there is a gap in the available information, an additional survey or case study may be needed. In the case of local, village-based activities, a participatory problem identification maybe more appropriate. The options for possible second steps are:

Case studies. These are in-depth studies of a number of farmers or households, e.g. observing farmers and their families over an entire season or several seasons, with frequent visits and participation in planting, harvesting and seed selection activities.

Surveys, PRAs and RRAs. An extended survey to collect more information is also called a 'diagnostic survey' (see following section). Participatory or rapid rural rapid appraisal (PRA or RRA) are considered to be alternatives to formal surveys in which farmers or farmer groups participate in the generation of information. They can involve group discussion, community or transect walks, mapping of the farm or village, preference ranking (see Box 11.3).

Participatory experimentation (see Chapter 12). Participatory experimentation follows the same process as other forms of experimentation. To ensure an effective involvement of the farmers, some other tools and methodologies are used to define the objectives, to plan, implement and evaluate the experiments. A good relationship with the collaborating farmers has to be established before the actual process of experimentation can start. Community or transect walks, resource mapping (Box 11.3), the discussion of secondary data, etc. can help to start building such a relationship and at the same time to identify constraints.

Box 11.3 Participating farmers, participatory appraisals and group evaluations

PRA and RRA
Participatory Rural Appraisal and Rapid Rural Appraisal as tools for information gathering emerged as a reaction to the feeling that conventional surveys did not allow for sufficient input from the target group in the assessment of the situation. PRAs or RRAs use group discussion, community or transect walks, mapping of the farm or village, and preference ranking to generate the information for problem diagnoses and priority setting. PRA and RRA are supposed to be time-efficient methods to collect the required information. Experience suggests that when properly planned and organized PRA and RRA also demand considerable efforts: rapport with the farmers has to be established, group participants selected and group meetings organized.

Group evaluation
Group evaluations are most productive and efficient in exploratory stages of research when farmers' criteria, and concepts for decision-making are not very clear. Group evaluations can also be used to evaluate a large number of alternatives with farmers, especially when this represents a tedious and exhausting task for an individual. Finally, group evaluations are useful for providing feedback to farmers about the results of previous trials or evaluations, to obtain their interpretation of these results. In such group discussions, the researcher/technician is principally a facilitator; stimulating the discussion by asking. The facilitator should be assisted by a reporter.

Who participates?
The composition of a group is very important. The farmers should be interested and there should be trust among the partners. The participating farmers should represent the target group. Discussion and dissent within a group can be very fruitful for the researcher-technician to get an overview of the diversity of opinions held by farmers and to understand the differences in objectives or resources. Some diversity in the group allows for different opinions and experiences to emerge. On the other hand, homogeneity and shared objectives and interest of the participants are needed to make the evaluation valuable. For instance, the participants in the evaluation of maize varieties for milling should be women for whom the preparation of the flour is a significant activity, otherwise the evaluation criteria will not be valid. Differences in social status may strongly affect group dynamics and inhibit people from speaking. When participants are expected to be very different, two or more groups may be formed, for example a men's and a women's group.

Advantages and disadvantages
In general, the following advantages and disadvantages of group discussions can be distinguished.

Advantages:
- o they stimulate discussion of evaluation criteria, especially when there are conflicting opinions;
- o they help to motivate the farmers and sustain interest in an evaluation;
- o they can be effective in terms of the researcher-technician's use of time;
- o they provide immediate feedback;
- o they provide an opportunity to involve farmers in the evaluations that are often under-represented in field trials.

Disadvantages:
- o they can be dominated by some members, leading to false consensus and misleading evaluation;
- o opinions will be withheld on sensitive subjects unlikely to be discussed openly;
- o they can be effective only if group activities are culturally acceptable;
- o they are less reliable for quantifying farmer preferences because members influence each other;
- o identifying or forming groups that represent user populations or fit research purposes may be logistically difficult, or time consuming;
- o farmers get tired of repetitive meetings.

Based on: Pretty *et al.*, 1995, and Ashby, 1990.

Events. These may include the organization of seed fairs, setting-up seed banks, or initiating co-operative seed production. The implementation of such events requires a similar process of formulating objectives, planning, implementation and evaluation. Examples of such action are presented in Chapter 2.

11.4 Surveying a local seed system

After the first orientation through informal interviews, consulting secondary information sources, and possibly an informal survey (Section 11.2), a diagnostic survey can be planned to collect more detailed, focused information on the local seed system. Such a detailed survey should provide the baseline data to be used in the final evaluation of the assistance. The informal survey, the interviews and information from secondary sources are used as an input in the definition and planning of a diagnostic survey. Depending on the objective of the survey, the focus might be on one or more of the following elements:

(1) *The use of varieties in selected crops:*
 – varieties used in the area;
 – farmers' use and knowledge of these varieties; and

170

– varietal mobility: introduction, adoption, loss, and the diffusion of varieties.
(2) *Seed sources and seed exchange:*
  – farmers' seed sources: own farm, neighbours/relatives, grain market, formal sources;
  – the availability, quality and price of these seeds;
  – seed exchange mechanisms, and flow of seeds between villages or regions;
  – access, availability, the adoption of seed and seed technologies; and
  – seed security problems.
(3) *Seed production, selection and storage:*
  – farmers' seed production, storage and seed selection practices;
  – farmers' knowledge of seed selection.
(4) *Gender analysis.*

The first three points are further elaborated in Box 11.4, the fourth is discussed in Section 11.7. The points mentioned in the box are used for the formulation of questions, such as for example in the sheet presented in Figure 11.2. It is again essential to formulate as precisely as possible what information is needed, and how it will be used, to prevent the questionnaire being too long. It is a good exercise to discuss this with the team and write down the answers to: (1) 'What do we want to know?' and (2) 'Why do we want to know it'. Too lengthy interviews are also a waste of the farmer's time, who may take the view: 'I give you all this information, and what do I get out of it?'

11.5 Designing seed system surveys

Stratification and sampling
Surveys that are meant to produce information or an inventory over a relatively large area involve a larger number of interviews, and choices have to be made as to which villages or farmers to include. To cope with the limited time availability a number of representative communities can be selected within the defined study area; a number of representative farmers within each community is then selected for interviewing. This is known as a stratified survey, with the strata being first the village and second the individual farmer. An advantage of such a stratified survey design is the fact that possible differences between farmers (large vs small, or men vs. women) and communities can be made identifiable. For example, two villages in the slightly drier area can be selected and compared with two villages in the area with more rainfall. From the groups of rich and poor farmers equal numbers of farmers can be selected. The orientation phase should provide the information on which to base assumptions

Box 11.4 Guidelines for seed system surveys

I. Farming system
o General information (name, sex, location, ethnic group, religion, wealth group, etc.)
o Information on the composition of the household (number of persons, number of persons working, children, migrated family members, hired labour, etc.)
o Labour (task men, women, elderly, children, hired labour)
o Ownership (owner, tenant, landless) and decision-taking (who decides on varieties, inputs, marketing)
o Economic situation (market sales, off-farm income, credit)
o Information sources (other farmers, co-operative, contact with agencies, extension service, radio, information leaflets, etc.
o Characterization of the farm (size, fertility, irrigation, gradient, variation in growing conditions, etc.[1])
o Livestock.

II. Cropping pattern and use of varieties
o Cropping pattern (plot sizes, crops and varieties, rotation)
o General agronomic practices (hand/machine sowing, quantity of seed used, plant density, harvesting)
o Use of varieties for different purposes (why are particular varieties planted, identification of important variety characteristics, comparison between varieties for different characters, the relative importance of the different characters such as taste, marketability, etc.)
o Knowledge of the agronomic behaviour of different varieties (disease resistance, conditions under which different varieties perform well, etc.)

III. Seed sources and seed exchange[2]
o Number of years the variety has been planted
o Original source
o Seed source last year, seed source under normal conditions, payment (friend, neighbour, market, seed programme)
o Seed replacement (frequency, reason for replacing own seed with better seed from others, e.g. diseased, degenerated genetically, etc.)
o Frequency of seed lost/not being able to save
o Reason for not being able to save (poverty, drought, etc.)
o Source of new varieties (actively searching, where, from whom obtaining seeds of new varieties)
o How are new varieties tried out?
o How are seeds of new varieties multiplied if small quantities are obtained (are they mixed with other seed lots, kept separate in separate fields or near the house)?
o Payment for new seeds (loan, ratio for swapping the seed, access, i.e. similar price for everybody, etc.)
o Distance to different seed sources

○ Who supplies seeds to others (farmer him/herself, identification of recognized supplier in the village for seeds of new varieties, for high quality seed, is there always seed available, frequency, payment)
○ Government interventions in seed supply and its effect on what farmers are planting.

IV. Seed production, selection and storage
○ Seed selection (in the field, after harvest, during storage, before planting)
○ Use of special inputs in the field (extra fertilizer, crop protection chemicals)
○ Special activities in field for seed (using special field, marking a special part of the field during the growing season, roguing weeds or off-type plants, marking special plants)
○ Storage place of the seed (together with consumption produce, other room/kitchen, air-tight containers, bags, in bulk or ear/cob, etc.)
○ Special conservation treatments (drying before storage, removing diseased seeds, use of chemicals, ash, or herbs, exposure to light, etc.)

Notes 1. Not all these factors may be relevant or require detailed information.
2. Each variety may be different. It may be simplest to ask if there are any exceptions or whether answers are similar for all varieties.

regarding such differences. These assumptions are the basis of hypotheses which can be tested in the survey.

The survey design should ensure that the farmers or households within each group are relatively homogeneous. The number of farmers or households selected per group does not need to be very high. A diagnostic survey with six or fewer groups and 10–25 interviews per group may provide sufficient information. Surveys with more than 100–150 interviews become huge and costly operations, often not justifying the costs. If it is felt that more interviews are necessary, the objectives of the survey may have to be reconsidered.

In the site selection there may be a bias towards selecting the more accessible villages or areas, to save time and transportation costs. Such bias may, however, undermine the representativeness of the collected information.

If there is to be random sampling of farmers in a particular area then it may be possible to select every fourth or fifth house in the village or along the road; if nobody is found there, then the neighbouring house is selected.

Before the start
Before the survey starts, the community leaders and village authorities should be informed about the data and purpose of the survey. Similarly, before each individual or group interview the purpose of the interview should be explained.

I. FARMING SYSTEM

1. Farmer's name ....
2. Sex (M/F) ....
3. Age class ....
4. Location ....

5. Size of the farm:
   1 - total area ....
   2 - cultivated area ....

6. Ownership of the farm:
   1 = owner   2 = tenant   3 = tenant/owner

7. Number of persons being part of the household:
   1 - number of economically active adults ....
   2 - number of children ....
   3 - number of senior family members ....
      (not working in the field)

8. Use of external labour
   1 = no
   2 = exchange labour with neighbours
   3 = labour in exchange for seeds, grain, food
   4 = hire payed labour
   5 = combination of other ....

9. General characterisation of soil fertility
   1 = favourable
   2 = intermediate, reason ....
   3 = marginal, reason ....

10. Irrigation
    1 = yes
    2 = no
    3 = part of the farm

11. Most important information source
    1 = relatives, friends, community
    2 = extension service
    3 = NGO
    4 = radio
    5 = leaflets, paper
    6 = other ....

Figure 11.2 Examples of points for a questionnaire for survey of seed and variety use.

174

II. CROPPING PATTERN AND USE OF VARIETIES
(based on a map drawn with the farmer, for example Figure 11.3)

Farm	Plot size (ha)	Crop: variety	Seed source 1999	Grown since 19....	Original seed source
Plot 1	0.15	Beans: Carioca	1	95	A:4 B:2
Plot 2	0.40	Maize: Dekalb	3 and 1	96/90	
Plot 3	0.25	Maize/beans: H-15/			
Plot 4	0.20	Jalapa			

Notes:
1 = farmer's own; 2 = friend or relative; 3 = other farmer; 4 = market; 5 = seed project/ government
A: 1 = friend or relative; 2 = other farmer; 3 = market; 4 = seed project/government
B: 1 = local; 2 = outside

12. Variety characteristics
(Farmers' comments on the characteristics and use of different varieties, using neutral, non-leading questions like: why do you plant this variety?; what do you use the grain for?; etc.)
– agronomic characteristic   ....
– storage behaviour   ....
   – keeping quality   ....
   – vulnerability to pests   ....
   – dormancy   ....
   – seed quality   ....
– consumption for   ....
   – special occasions   ....
   – market   ....
   – other purpose   ....
– culinary use and quality   ....
– need for agro-chemicals   ....
– market price and whether fluctuating   ....
– other comments   ....

13. Most important variety characteristics:
   1 - ....
   2 - ....
   3 - ....
   4 - ....

14. Best, second best, third variety in relation to (ranking):
   – drought resistance 1)...........2)............3)...........
   – consumption quality 1)...........2)............3)...........
   – cooking time 1)...........2)............3)...........

15. Newly introduced varieties   ....

16. Lost varieties still remembered

Name	Characteristics	Reason for disappearance
1 -	....	....
2 -	....	....
3 -	....	....
4 -	....	....

III. SEED SOURCES AND EXCHANGE

17. CROP A ....

18. CROP B .... (data vary per crop and can be different per variety)

19. Normally used source of seed:
   1 - farmer's own
   2 - friend or relative in the community
   3 - other farmer in the community
   4 - other farmer in other location
   5 - market
   6 - seed project/government

20. Distance from the off-farm non-local source
   1 - local
   2 - other community <20 km ....(name)
   3 - other community >20 km ....(name)

21. Using seed from other sources:
   1 - every year
   2 - regularly, every .... years
   3 - only in case of calamities
   4 - never
   5 - other ....

22. Reason for regularly using seed from other sources
   1 - reasons of seed quality, other source is better
   2 - variety changes, degenerates, or 'gets tired'
   3 - sold out in years when grain prices are high
   4 - eating or selling the seed for cash needs

   comments on 21 & 22:
   ....
   ....
   ....

23. Reason for using seed from other sources in case of calamities
   1 - crop loss: climate (drought)
   2 - crop loss: crop pest or disease
   3 - sickness family member/funeral/wedding/other social
       commitment
   4 - other ....
   comments ....

24. Source of seed in case of calamities:
    1 - farmer's own
    2 - friend or relative in the community
    3 - other farmer in the community
    4 - other farmer in other location
    5 - market
    6 - seed project/government

25. Seed exchanged for:
    1 - cash
    2 - swapping for seed/grain of other variety
    3 - labour
    4 - share cropping
    5 - other ....

26. Number of farmers who were supplied seed to last planting ....

27. Seed exchanged for
    1 - cash
    2 - swapping for seed/grain of other variety
    3 - labour
    4 - share cropping
    5 - other ....

28. 'Price' of the seed   ....

29. Recognized seed farmers in the community
    1 - rich: always has seed
    2 - rich: has good quality
    3 - poor: always has the variety
    4 - poor: has good quality
    5 - other: ....
    comments:

30. Seed sources for local bean varieties

31. Regular seed sources for improved bean varieties

IV. SEED PRODUCTION AND STORAGE

32. Field production practice (e.g. extra input, roguing, other measurements)

33. Practices of seed selection of farmers' own saved seed
    1 - selection of pods/ears/plants in the field before harvest
    2 - selection of pods/ears/plants during harvest
    3 - selection of seeds/pods/ears during storage
    4 - use of left-over seeds

34. Characteristics selected for:
    ....

35. Selection of seeds from other varieties
    1 - similar   2 - different

36. Storage of seeds: storage in bulk with consumption grain, stored separately, other room or kitchen)

It is useful to have a trial run of a number of dummy interviews to test the checklist, the questionnaires, the options on the questionnaires and the use of data sheets. The trial also serves to monitor the team's interaction with the farmer.

Team composition
If the survey questions relate to a specific area of knowledge (e.g. seed potatoes) it may be worth inviting a person with the relevant expertise (knowledge of potato production) to assist in the survey.

In many cultures it is customary that the man, being the head of the household, does the talking, while the rest of the family listens. When the interviewing team includes at least one woman, women in the household will speak more freely and better conditions are created for the collection of household data, e.g. regarding the household garden (flowers, herbs, medicinal plants) or the kitchen (food processing, cooking consumption quality).

Conversation and communication
Having a free-flowing conversation during the interview is one of the basic conditions for an accurate survey. The farmer should be encouraged to give objective answers, without being distracted by an interviewer who appears to be more interested in filling in the questionnaire. There is a delicate balance between addressing the farmers and completing the survey form. This balance is influenced by using open or closed questions (see Box 11.2)

Information collection
During the interview a simple data sheet for each variety may be used. Collecting information on each variety, however, easily results in a lengthy and boring inventory, detracting from the discussion of issues raised in open questions and of the farmers' interest. A short cut may be to ask a farmer whether seed sources, selection practices or other aspects are similar for all his or her other varieties of the same crop. Differences are most likely to occur between local and improved varieties, between varieties grown for home consumption and the market, and between open pollinated and hybrid varieties. The survey may not cover all the crops, nor all aspects of seed handling; this depends on the objectives of the survey.

Mapping the farm with the crops and varieties in the different plots is an effective tool for an inventory of the varieties grown by farmers (diagramming and visualization, see also Box 11.3 and Figure 11.3). When only asking the farmer which crops or varieties are grown, the ones occupying tiny areas, along the borders or in the corner of a field (the ones with particular uses, e.g. for special dishes and brews) are easily omitted. This information can be of particular interest for the survey, while the farmer may be used to an interviewer who is only interested in the 'important' varieties, i.e. those occupying most of the land, or those recently introduced. Valuable plants with household uses may be present in the household garden, among the weeds or on the land used for grazing.

Accurate yield data are difficult to collect through a survey. Farmers may recall the 'very good' years and 'very bad' years, but are not likely to remember the exact number of bags or donkey loads they harvested. The data may lack the required accuracy and are not very useful for calculations. Also, if only some farmers do recall yields of earlier years and others do not, the value of the information will be limited. One option may be to collect the information on the last growing season only. This may not cover variations between years, but allows fair comparison between varieties, fields and farmers.

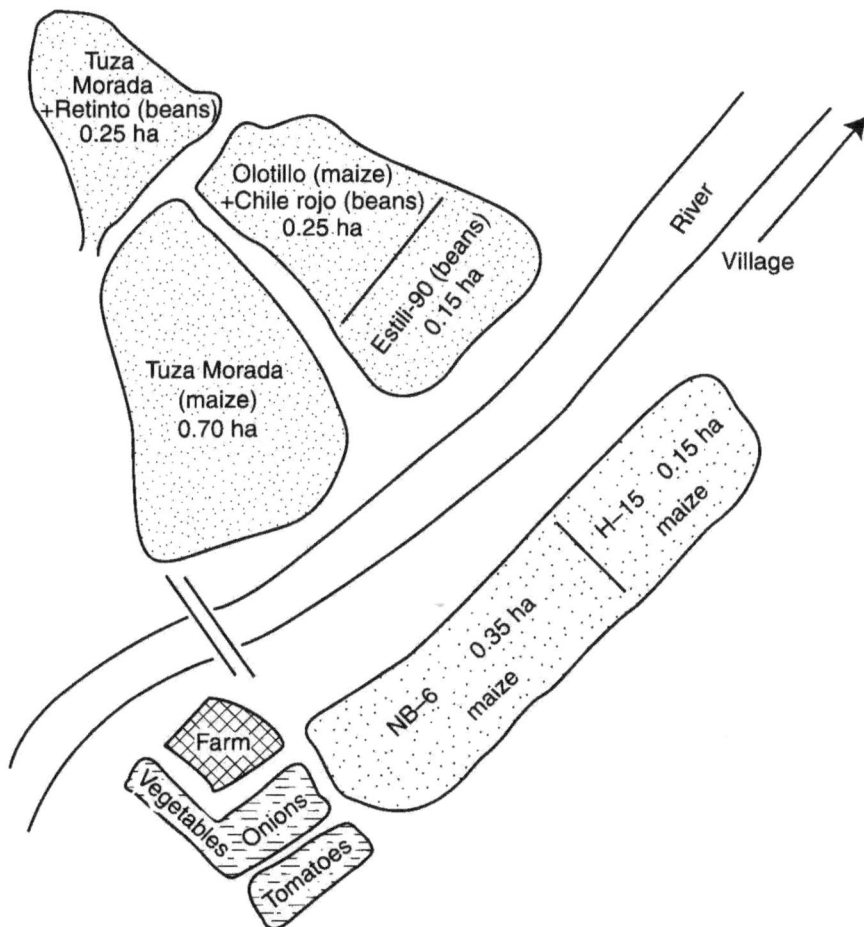

Figure 11.3 Diagram of a farm and cropping pattern in Nicaragua, as may be drawn up during an interview with a farmer.

For other types of data it may be possible to compare the present situation in two different places, for example where access to improved varieties varies. Comparison of two different points in space are not similar to comparison of two different points in time, but may yield more accurate data.

11.6 Preparation of the questionnaire and data collection sheet

There are three type of questions in a questionnaire: those answered with yes or no, those in which a number of options are indicated (including an open one for answers not previously considered), and open questions. The first two are the easiest to process. The last one allows greater freedom for the farmer to tell his or her story, but these answers may be difficult to analyse, and are often not processed for the survey report. With this type of question it must therefore be clear what sort of information is required. True stories or experiences may be interesting, for instance to be used for the structuring of a future survey, but should be restricted to a limited number. An informal survey in a first orientation phase will have more open questions; a wider survey with more interviews and based on preliminary information already collected will have more questions of the first two types.

Questionnaires are usually structured as a guideline for the conversation; they do not necessarily follow the a systematic sequence of data for data processing. Simple, short questionnaires may be filled in while talking to the farmer. In reality, the conversation during the interview is more free-flowing and often deviates from the sequence of questions in the questionnaire. Instead of having to interrupt the farmer constantly, a second team member may take prime responsibility for recording the information.

After the interviews, the information is entered on to a processing sheet. From the processing sheet the numbers can easily be entered into a database program. For example, the processing sheet in Figure 11.4 is a printout of a spread sheet of a database program like DBase, Excel or Access. It is also possible to enter the data directly from the questionnaire into the database program. In this case, the questionnaire and database structure should be very similar to avoid mistakes, bearing in mind that the person who enters the data often did not participate in the interviews.

Thinking ahead to the design of the processing sheet and analysis of the data, forces the researchers to be clear about the type of answer expected and how they will be processed, including yes or no questions, a number of optional answers, descriptive information and comments. The preparation of such data sheets should be directly linked to a database structure.

Farmer (no.)	Crop 1 = maize 2 = beans	Variety	Var. code	Plot size (ha)	Seeds available on time	Origin of variety	Source	Renewal	How often	Orig. seed source	First year	Yield	Pest resist.	Taste	Storage
1	1	tusa morada	141	0.35	True	1	1	False	0	1	87	1	1	1	1
1	2	revolucion-84	245	0.40	False	1	7	False	0	2	90	1	1	1	1
1	1	rocamex	135	0.25	True	1	1	False	0	1	89	2	1	1	1
2	1	rocamex	135	0.15	True	9	1	False	0	1	70	2	3	1	1
2	2	chile claro	207	0.40	True	9	1	False	0	6	87	3	3	1	2
3	1	rocamex	135	0.25	True	3	1	False	0	1	89	1	2	1	1
3	2	revolucion-84	245	0.25	True	2	1	False	0	2	89	1	2	1	1
4	2	esteli-150a	220	0.15	True	9	7	False	0	2	91	1	1	9	9
4	1	rocamex	135	0.50	True	1	1	False	0	6	70	1	3	1	1
5	2	chorotega	215	0.10	False	1	1	False	0	1	87	1	1	1	1
5	2	revolucion 84	246	0.55	True	1	1	False	0	2	89	1	1	1	2
5	1	rocamex	135	0.34	True	1	1	False	0	2	60	1	1	1	3
5	1	nb-6	131	0.40	False	1	1	False	0	2	86	1	1	1	3
6	2	h-15	0	0.20	True	1	1	False	0	1	85	1	2	1	1
6	2	balin	201	0.27	True	1	1	False	0	1	89	1	1	1	3
6	1	maizon	126	0.12	True	1	1	False	0	1	60	2	1	1	1

Figure 11.4 Completed spread sheet with information from the data sheet. Depending on the computer program used for processing the information, 'yes' and 'no' can be coded as true and false or as 1 and 2; different varieties may be assigned a code number. Valuable computer programs for analysis are SPSS and Excel.

Box 11.5 Gender and genetic resources

In local seed systems genetic resources are used, managed and maintained by farming communities and individual farming households. A household is not an undifferentiated group of people, however, and to target the needs of a household means to understand the very different needs of its members.

Men and women together comprise families and farms, each with their own role to play in daily life. While an individual's biological sex is simply a product of being born male or female, what one is socially expected to do as a man or a woman is learned behaviour. In other words, gender, although associated with biological sex, is learned. Human beings learn what is male work and female work, male roles and female roles. These roles are accompanied by rights as well as responsibilities. Thus, gender becomes a factor that must be considered in the management of the farm and farm resources, one that varies with the culture, crop, and the socio-economic and biophysical production environment. Other differences in tasks and responsibilities, such as those of the young and old, or rich and poor are not further dealt with here, but must also not be overlooked when targeting interventions to improve seed supply (see also Box 2.6).

Gender, varieties and seeds in the local system
In many cultures women are responsible for breeding, selecting, diversifying, and maintaining the diversity of the gene pool. Children also play a role. Native Central American women select maize ears for seeds as they shell maize for consumption. Among the Ifugao in the Philippines, women are responsible for selecting the rice ears from the field as seed for the next planting. In Rwanda and Malawi, women manage bean seed; and in the Andes women are responsible for separating out roots and tubers, deciding which are to be eaten, which stored, sold or bartered, and which are to be used as seed for the next planting (see also Box 2.1).

Kpelle women in Liberia maintain over 100 varieties of rice in swidden cultivation. Young girls become encultured to their gender roles as the managers of rice diversity and memorize information about the varieties and the micro-environments to which particular varieties are suited. In Malawi, young boys play with beans pilfered from their mothers' storage bins, and this results in bean seed being traded within and between villages unintentionally (Martin and Adams, 1987). The boy who wins the beans over the other boys either returns them secretively to his mother's storage bin or sells them at the market for pocket money.

Gender-specific crops
In most cases men and women have different roles in agricultural production, and also contrasting reponsibilities for different crops or varieties. In many places, men have the responsibility for cash crops (to which most external inputs may be allocated) and women for the food crops or varieties serving home consumption. In Africa, men may grow modern vegetable

crops such as cabbage and tomato, for the market, while women manage the indigenous vegetable plants. Highlanders of Papua New Guinea divide crops into those that men can plant and tend (sugar cane, bananas and yams), those that only women cultivate (sweet potato, cucurbits and selected greens) and those that are neither male nor female specific. A man who plants a woman's crop risks social ridicule; a woman who plants a man's crop may even suffer physical violence.

Gender-based varietal preferences and technological change
Men and women often have different varietal preferences, though sometimes these can change, as noted by Johannessen (1982). Years ago, women preferred the softer maize grains: it was their responsibility to hand-grind the corn, and it took less soaking with lye and a shorter cooking time if the maize was softer. Men concern themselves primarily with the keeping qualities of maize, and therefore men favour flint starch – *crystalina* – which is considered to be most resistant to insect attack. Diesel-powered maize-grinding mills have now been introduced, however, and this has reduced women's concern for the hardness of the flint corn.

Plant breeding, seed programmes and extension
Agricultural researchers and extensionists have neglected the different roles of men and women in the past, and in particular the role and interest of women. The most obvious omissions in plant breeding have been the assessment of characteristics that women deem important to the crop for domestic purposes, including taste and cooking quality.

Women are active innovators and yet remain dependent upon informal contacts rather than institutional sources for information and seed. A study in Sudan showed that women adopted new seed varieties, but had to get them from informal sources, such as the market, family, or other villagers. All agricultural extension field agents were male, and extended technologies only to men.

When gender-specific expertise is recognized and incorporated into research and extension, the results can be very successful. The bean varieties that the women farmers in Rwanda chose for on-farm testing had higher yields than varieties breeders chose, while at the same time satisfying a range of additional selection criteria (see Box 2.3).

Incorporating gender into research and development
The best way to ensure that the needs of both male and female farmers are accurately targeted is to build in a gender component early to any endeavour to understand or enhance the seed supply.

Gender analysis includes the study of who does what (activities analysis), who has access to what (resource analysis) and who benefits (benefits/ incentives analysis). With insights into the different roles and responsibilities it is possible to target the needs better and to predict the impact of planned intervention.

11.7 Gender analysis

An analysis of men's and women's activities, resource access and benefits/incentives can be an integral part of a seed survey (see Box 11.5). A thorough gender analysis may be time consuming, however, and should be carried out as a separate study, preferably with a limited number of households.

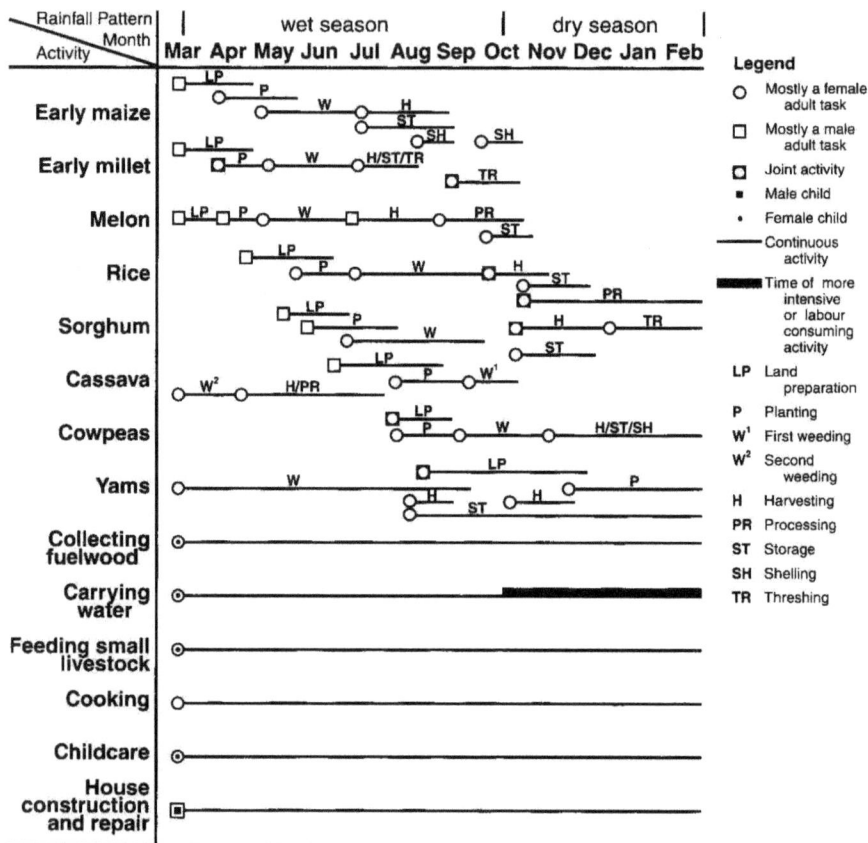

Figure 11.5 A cropping calendar showing the tasks of women and men for different crops (from Feldstein and Poats, 1990, based on Burfisher and Horenstein, 1985).

Activity analysis
Suggestions for improvement are often related to shifts in labour demand. Disaggregation of the different tasks in the household and production into tasks of men and women, of the youngsters and old show where labour shortages exist and what implications suggested activities will have for each of them. The information can be used to construct a cropping calender, as

184

in Figure 11.5. An analysis per variety is very laborious; an analysis per type of variety is likely to be more effective and may yield interesting differences. For instance, local vs. improved varieties, hybrids vs. OPVs. For a brief, sketchy analysis of the gender aspects, a worksheet such as in Figure 11.6 can be used. Filling out such a worksheet will demonstrate that activities such as seed selection and storage are closely related to access to seed as a resource and knowledge. The worksheet can, for instance, be useful to question what the effect would be of introducing a hybrid variety, or at whom improved seed selection should be targeted (see Box 11.5). A general framework for gender analysis in farming systems can be found in Feldstein and Poats (1990).

	Activities F	M	Resources F	M	Incentives F	M	Costs F	M	Benefits F	M
Local 'Blanca'										
sowing	★	-	⇑	-	Control over	-	⇑	-		-
weeding	★	-	Planting	-	production	-		-	Collection	-
harvesting	★	★	material	-	for household	-		-	of weed	-
seed selection	★	-	knowledge	-	consumption.	-		-	for use as	-
seed storage	★	-	⇓	-	Taste, cooking quality.	-	Labour	-	vegetables	-
use of produce	-	-	-	-	Brewing quality.	-		-	Household food	-
use of secondary					Seed for exchange.				Collection weeds	-
produce	-	-	-	-	FOOD + SEED	-		-	Animal feed	
use of seeds	-	-	-	-	SECURITY	-	⇓	-	(stems)	-
Hybrid '101'										
sowing	★	-	-	Cash	-	-	-	Cash	-	-
weeding	-	-	-	needed	-	-	-	-	-	-
harvesting	★	★	-	for	-	-	-	-	-	-
seed selection	-	-	-	seed + input	-	-	-	-	-	-
seed storage	-	-	-	purchase	-	-	-	-	-	-
use of produce	-	-	-	-	-	Cash	-	-	-	Cash
use of secondary									Animal feed	
produce	-	-	-	-	-	-	-	-	(stems)	-
use of seeds	-	-	-	-	-	-	-	-	-	-

Figure 11.6 Worksheet for gender analysis of seed-related activities. Note: Possible differences between the use of the local variety 'blanca' and a hybrid variety '101' are indicated. The hybrid maize is recommended to be grown with use of herbicides, and the produce cannot be stored, seed is purchased yearly. The local variety is used for tortillas and brewing *chicha*.

Resources analysis

Because decisions depend upon it, it is important to know who has control of critical resources the power to decide on how the resource is used. This may include whether grain is used for household consumption, sold, or kept for seed, but also, who has access to the land and capital (including credit), and who can influence water availability through irrigation. Knowledge is a resource which is often overlooked and which may be very important in relation to seed selection practices.

Rio Tinto – Maize: 8 varieties planted

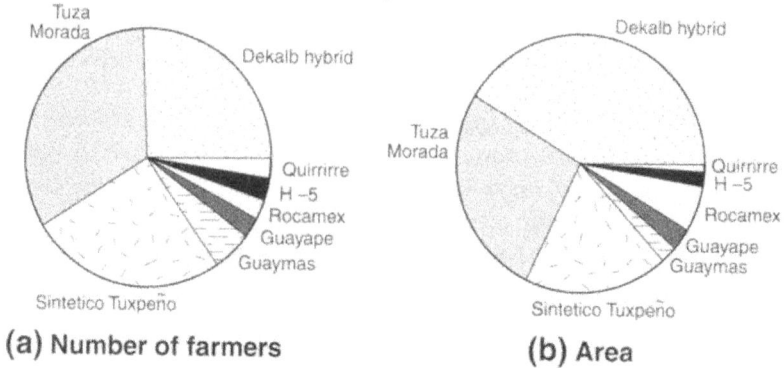

(a) Number of farmers

(b) Area

Rio Tinto – Beans: 5 varieties planted

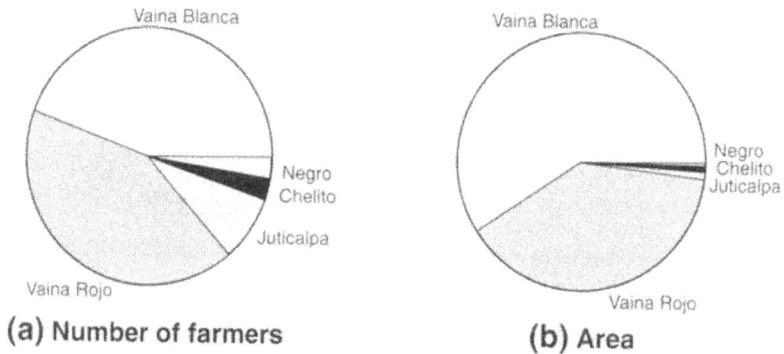

(a) Number of farmers

(b) Area

Figure 11.7 Example of ways to present survey results. Number of farmers planting different maize and bean varieties and the area planted in 1991 (as a percentage of the total), Rio Tinto, Honduras.

Benefits/incentives analysis

The analysis of benefits particularly refers to who has access to or control of the output of production, i.e who is the user of the output or receives the benefits of it. This refers to the crop produce, but can also include the collection of weeds in the crop to be used as a vegetable in household consumption or sale. Another benefit of the crop may be the opportunity to graze livestock on the crop stubble. The next step in an analysis is to determine what are the characteristics of the output that can be considered an incentive for the user, such as for instance the good cooking characteristics and taste of a variety. This information helps in the analysis of possible changes in the character of the output or shifts in benefits (e.g. more milk per cow, a cash crop instead of grazing grounds, a cash crop instead of a local variety for brewing beer). When a suggested improvement involves a change in labour demands,

186

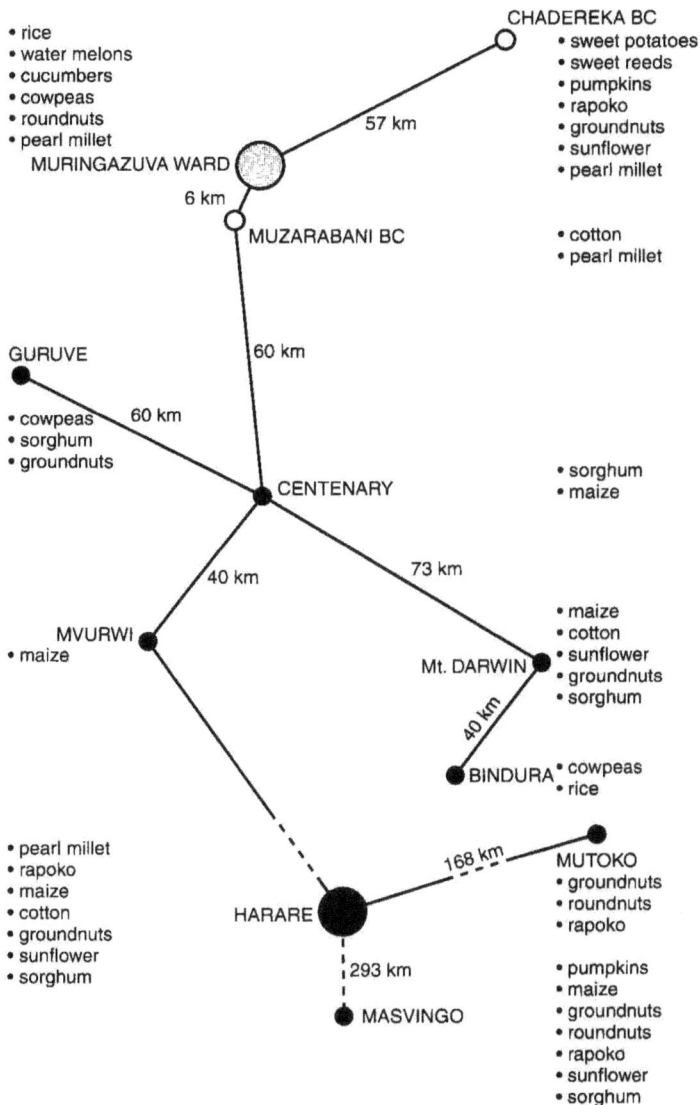

Figure 11.8 Example of how information on seed sources can be made visible: a mobility map from Muringazuva ward in Muzarabani district showing distances covered by farmers in their seed-sourcing activities (Oosterhout, 1996)

then this change should result in corresponding benefits: the ones expected to do more work should also be important beneficiaries.

Inclusion

Inclusion principally refers to an analysis of who is included in the planned activity (e.g. are women included, do the poor participate?), what their role is (e.g. consultant, decision-maker, enumerator, interviewer) why they are included (criteria) and how (e.g. individual interviews, group discussion, experimentation committee). The purpose of the analysis is to make sure that all significant participants are actually included, in a way that corresponds to their role and responsibility.

Information on gender is collected so that the plan of action will be more effective and better targetted. The worksheets, maps or profiles help to define which technology (varieties, seed production, storage or selection practices, specialization, etc.) may be evaluated or improved. Gender analysis is an additional tool in the planning of participation. The information can be collected via individual interviews or group discussions, and can be used to refine the participation in research and development action.

11.8 Presentation of results from a survey

Results should always be presented to make the information accessible and comprehensible. A table full of statistical analysis may not tell the researcher very much about the situation, whereas relationships may suddenly become clear when the information is presented in a 'user-friendly' way, such as graphs or pie-charts (see Figures 11.7 and 11.8).

12. Participatory experimentation

12.1 Priority setting and planning

12.1.1 Diagnosis

Farmers and communities may not yet be interested in seed and variety-oriented activities. Even if seed quality or varietal adaptation are considered problems, for example, soil erosion or access to credit may need to be addressed more urgently.

If this is the case, the work on seeds may have to be cancelled, postponed, or continued with a smaller number of farmers who show a particular interest in seed issues. Methods such as those described in Box 11.3 may be used both to prioritize different constraints and analyse the opportunities for solving them.

When farmers are interested in experimenting with seeds and varieties, the decision on 'things to try' requires careful and a more accurate definition, and prioritizing of constraints. Investing time and energy in this stage will definitely pay back. It allows for a more focused search for alternatives, and a better planning of treatments and comparisons in the experiments.

When the options for solving the highest priority constraints (e.g. credit, or marketing options) are beyond the reach of both the community and the outsider, working on seed problems can be used to stimulate community action, and may even lead to opportunities to access credit.

12.1.2 Planning the experiments

Once problems have been prioritized, possible solutions need to be listed, based on the combined knowledge of farmers and outsiders. (Box 12.2). What follows is suggested experimentation with possible solutions.

The design of the experiments involves a number of aspects to be discussed and decided with the farmers:

o Which treatments are to be tried (varieties, storage conditions, etc.)?
o What should be the lay-out of the trial (field, stores, etc.), which will depend on whether or not a statistical analysis is needed at the end?
o What observations will be recorded, and who will the observer be?
o How will the data be processed?

It is also important to determine beforehand whether the owners of the fields, stores, and the persons involved in the execution of the trial will be paid for their input, and how.

Box 12.1 Participatory approaches for ranking and evaluation

Pair-wise comparison. With pair-wise comparison each alternative, i.e. treatment or variety, can be judged better or worse than another, while reasons for this judgement are given. This technique rapidly becomes tedious if more than six items are being compared, so that it is best used once a smaller number of alternatives has been identified.

Complete pair-wise comparison can be carried out on three or four alternatives for example, in the following way: all alternatives are compared with each other: A with B, A with C, A with D; B with C, B with D; C with D. Treatments in a trial can be given simple names, for example, or symbols can be used to represent each treatment, and then shown to the farmer in pairs. Thus, pair-wise comparison can also be used to ask farmers to rank criteria. It is not wise to force choices without obtaining an understanding of any reluctance or difficulty the farmer may have in making a judgement between two alternatives.

Matrix ranking. The researcher can get additional insight into a farmer's criteria by asking him to rank several treatments with respect to specific criteria which have been identified previously. This technique, called matrix ranking or grid ranking, is illustrated in the table below.

The interviewer can ask the farmer to rank the four best-liked varieties with respect to yield, growth habit, disease resistance, etc. The interviewer begins with the question: 'Which of the four varieties you have selected is best with respect to yield? Which would you put in second place? . . . (third and fourth place)'. The ranking is repeated with respect to each criterion of interest. Other procedures of matrix ranking are also possible.

Matrix scoring method*, used by eight farmers to evaluate potato clones at Chullchunqani, Bolivia in 1994–95 (Thiele, et al., 1997)

Clones	Market	Wide eyes	Yield	Health-iness	Shape of tuber	Taste	Total
385240.2	24	21	24	26	22	8	125
720049	20	20	19	27	19	16	121
676008	22	24	23	14	17	15	115
575045	18	12	17	17	15	8	87
382171.10	18	15	11	16	14	8	82
84–75-16	13	13	10	18	14	9	77

*) Farmers evaluated one criterion at a time, placing one (poor) to three (good) grains of maize by each pile of clones. All clones were boiled and evaluated for taste by the same procedure. Scores of different criteria (number of grains) were summed.

Diagramming. Diagrams and visualizations are pictorial or symbolic representations of information. Visual methods allow non-literate and literate people to participate in a communication process as equals. They work

After the experiments are carried out, the results have to be evaluated and shared.

The remainder of this chapter presents some ideas on experimentation with varieties and seeds.

12.2 Experimentation with varieties: variety evaluation

12.2.1 Introduction

Evaluation of new varieties is the result of priority setting exercises (Box 12.1), especially where new characteristics are needed in a crop. This chapter concentrates on participatory evaluation of varieties, i.e. PVS (Participatory Varietal Selection).

PVS trials can take many different forms. At the one end of the spectrum is the trial designed and managed by the breeder or researcher-technician; at the other end are on-farm trials planned and managed entirely according to farmers' criteria. The choice of the location, the number and type of materials and the lay-out can vary according to the more specific objectives of the evaluation.

This chapter presents a general characterization of variety trials on the basis of their specific objectives. Recommendations on design and realization of a trial are related to these objectives. Many variety trials do not exactly fit any of the three categories presented below, but may have intermediate objectives and intermediate forms.

Box 12.2 Examples of options and suggested action for seed quality problems

Case 1. Poor germination of the seed

Possible causes	Suggested action
○ Untimely harvesting	Try early harvesting (in exceptional cases, later harvesting)
○ Poor drying	Build drying racks, local seed dryer
○ Rough threshing	Check seed moisture, introduce small seed threshers (group credit)
○ Poor storage conditions	Try cribs, polythene bags, community stores
○ Storage insects	Try local herbs, Actellic
○ Seed dormancy	Literature review

Case 2. Too many weeds, inert matter in the seed

Possible causes	Suggested action
○ Weeds in the field	Identify weed seed and link with field weeds
○ Harvesting technique	Heap heads on a tarpaulin (Striga), harvest infested fields by knife instead of sickle
○ Seed cleaning	Improve winnowing and sieving, introduce small equipment

Case 3. Diseases occur early in the season

Possible causes	Suggested action:
○ Seed transmitted?	Identify disease, check if seed transmitted
○ If completely infected	Find disease-free seed stock
○ If available	Treat seeds with chemicals
○ If small infection	Remove diseased plants in the field
○ If visible on the seed	Remove diseased seeds before planting

Case 4. Yields decline

Possible causes	Suggested action
If decline is seed related:	
○ Variety is mixed	Purification and maintenance (part B)
○ Variety unadapted	Participatory Variety Selection, or Plant Breeding

12.2.2 Different types of variety trial

For breeders and national researchers, variety selection is a process which narrows down a large number of materials to a smaller number of superior varieties. Normally, the first trials are for orientation and are carried out on-station. As the selection process proceeds with multilocational community trials, the number of materials in the trials decreases, and farmers' involvement increases, culminating with on-farm confirmation trials (see Figure 12.1).

	Characteristics	Objectives	Related activities
orientation trials	• on station • reseacher managed • large number of materials • small plots • repetition less relevant	• orientation on performance of materials • orientation on farmers criteria • 'shopping' for interesting materials	• selection of promising mate • farmers invited to field days
evaluation trials	• in the farmers' environment • collaborative management • 8–20 varieties • plots of 2–3m² or more • 2–4 repetitions • data collection & evaluation	• evaluation of performance in farmers' conditions • evaluation of variety characteristics in relation to farmers' criteria • multiplication of materials	• data collection • group evaluation (ranking, • selection of most promising
confirmation–adoption trial	• on farm • farmer managed • 1–3 varieties • large plots • repetitions less relevant	• acceptance or rejection of promising materials • multiplication of materials	• follow-up by reseacher • visits by neighbouring farm

reduction of no. materials
increasing plot size
increased farmers' involvement

Figure 12.1   Characteristics and objectives of orientation, evaluation and confirmation/adoption trials. G m orientation to confirmation/adoption, the number of materials reduces, and the plot size and the far rticipation increase.

Participatory aspects can be brought in at any of these levels. On-station selection trials can become on-station demonstration and participatory selection trials. Multilocational trials can become (researcher-designed and) community-managed trials, and the confirmation trials may be farmer-designed farmer-managed on-farm trials.

On-station demonstration trials
An on-station demonstration trial, designed and managed by researcher-technicians can include a large number of materials. Such a trial is usually meant for a first orientation; plot size can be small and replication may not be necessary. Special field days may be organized at which farmers are invited to evaluate the material. Farmers may be given coloured ribbons, or sticks to mark their favoured selections. Afterwards, the results of the day may be presented and discussed with the farmers. This is a good option both for early stages of a breeding-selection programme or for the use of a germplasm collection, and for when there is little information on the preferences of farmers.

Grouping of varieties according to maturity period and plant height makes evaluations more effective both for the breeder and for farmers. In large demonstration trials, local varieties should be planted on several plots in the trail in order to have a nearby reference at any time (Figure 12.2).

Farmers' interest can be stimulated when there is an opportunity for the farmers to take some seed of selected materials; the selected materials are then fed into the next phase of selection. Follow-up on what happens to these particular materials, and whether they are later exchanged with other farmers can provide very useful information. The breeder's interest is in obtaining a clear picture of the farmers' selection criteria, which should be included in future breeding objectives.

Community trials
When there is a good idea of what kind of materials are of interest to the farmers, selected materials are subjected to more serious evaluation at different locations. The variety evaluation trials could be moved away from the control of the researchers; trials with selected materials are situated in the farmers' fields and managed by farmers. These trials can take the form of community trials, i.e. trials in which a group of farmers manage and evaluate the experiment.

As the number of varieties included in such trials decreases, the plots may become larger. For quantitative evaluation of traits, replications may have to be included. Data can be taken on early plant development, growth habit, flowering time, disease incidence, yield and product quality. Group evaluations can be organized to discuss the observations. Community variety trials may be implemented on the farm of an individual farmer, in a communal field, a local school garden or any other site that people have

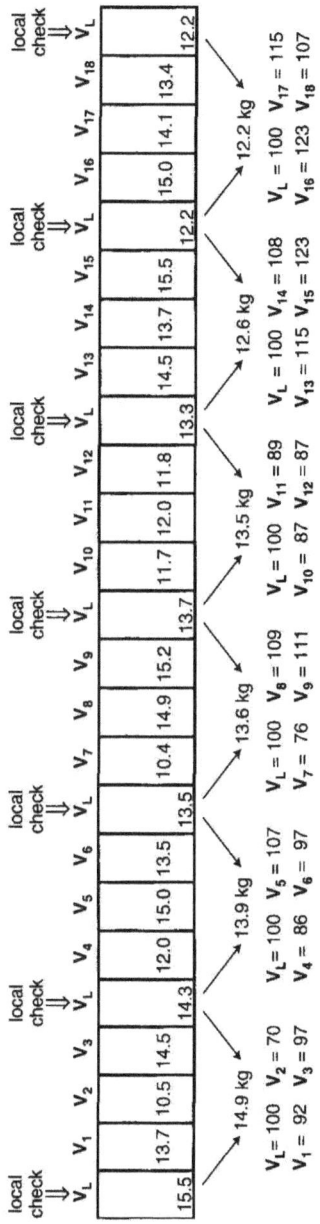

Figure 12.2 In large demonstration trials, planting a plot with the local variety every four plots allows for comparisons, even if the field is not homogeneous. For instance, yields can all be indexed, with the local variety being assigned the value 100: indices can be ranked (this procedure can also be used for repititions, whereby average indices of the varieties are ranked).

relatively easy access to. The trial may also be at the research station, but only if the farmers can determine the agronomic practices.

Wherever the trials are laid out, however, there is a risk that they will be treated with more care than common crops, e.g. with timely or extra weeding. This could create conditions which differ from the farms and result in sub-optimal selection. Where trials are laid out in communal plots with a wavering commitment from the participating farmers or community leaders, the opposite may happen. Poor land preparation and late weeding may spoil the trial.

This type of demonstration trials may have 8 – 25 entries. Selection for the following phase of evaluation can be done by the farmers or jointly by farmers and the researcher-technician or the breeder, who should be informed about the reasons for choosing the particular lines, in order to target priorities better in the future (feed-back of information).

On-farm trials

On-farm evaluation is usually carried out in the fields of individual farmers. Apart from obtaining further information on the value of the varieties in specific farming systems, such trials are also used for immediate 'rejection' or 'adoption' of the new material. The number of varieties in the evaluation has to be limited, since the farmer's key priority at this stage is to produce food and not to do research. Management and evaluation are to a large extent left to the farmers' criteria. Normally farmers do not replace their variety with a new one within one year. They usually plant a small plot with the introduced seeds and observe this crop carefully. Only when the introduction appears to be promising do they increase the proportion.

A farmer may be given one to eight varieties to compare with his or her own cultivar. The way the materials are planted, crops managed and varieties evaluated can be left completely to the farmer. Alternatively, a joint design and evaluation may be agreed upon. If desirable, a design which allows for statistical analysis may be laid out.

Since new varieties are planted in the farmers' crop production field, farmers must be warned that variety evaluations of cross-fertilizing crops carry the risk that poor characteristics may cross into the farmer's own variety.

Alternatives

Many intermediate forms exist in which the division of tasks between breeders and farmers differ from those just suggested. One alternative is to reverse the route described here: to start with a small number of collaborating farmers and a small number of new materials. When experience increases, the process may be scaled up to community level. When selection criteria are well defined, the farmers are interested and there is some level of organization for community experimentation, the formal breeding programmes and genebanks may be contacted to obtain materials or to discuss options for joint evaluation both on-station and on-farm.

12.2.3 Which materials to include

Which varieties to evaluate in community and on-farm trials depends on the needs and preferences of the farmers. Locally important varieties have to be included for comparison at any time. Normally, the choice of exotic materials (i.e. materials which are 'foreign' to the place) is limited to those varieties that are used by other communities in the region and varieties that can be obtained from the national research institutions and seed producers/importers. These seed companies may be able to supply international materials that have gone through an initial screening on-station for adaptation, or which have been tested in other parts of the country. They should also have a basic understanding of the potential for adaptation of the variety to the conditions in the locality and be able to indicate which materials may be of interest. The old OPVs from national programmes, if still maintained, may also be useful: these materials have often been discontinued because better-performing hybrids became available. Since OP varieties do not need a yearly purchase of seed, they may be preferred by many farmers. The regional germplasm collections of International Research Institutes are useful sources for materials, and requests for materials are commonly agreed to, even though the flow of seed may have to go through the national system and only small samples of seed per accession are provided. This can in some situations result in a serious bottleneck. Genebanks or other organizations maintaining germ plasm collections can only provide small samples of seed. They can be valuable for landrace materials which may have been lost by the farmers in parts of the country. NGOs or extension services in other parts of the country may be instrumental in providing both materials and relevant information.

The only way to confirm adaptation is to grow and evaluate seed in the region. Adaptation relates not only to the yield of the main product, but also to secondary products (e.g. straw), disease resistance, taste and quality, adaptation to cultivation practices such as intercropping, staking (beans), and so on.

Seed multiplication may be necessary to have sufficient seed for a trial. This has important consequences for the planning of activities, since trials may only be started one or two seasons later. It may be an option to start with varieties of which seed is more readily available and gain experience with evaluation trials, while building a relationship with the farmers. Other varieties may be included in later seasons.

The number of varieties

More varieties provide wider options, but evaluation of too many materials complicates the management and increases cost. Between eight and 25 different varieties can still be comfortably monitored and distinguished in one researcher- or community-managed experiment. If more varieties came out of the initial selection, different farmers may test different varieties; each farmer testing two or three alternative varieties together with his or her own. In such a set-up the new varieties may be planted as plots in the main production field. Farmers may visit each other's fields to discuss the merits of each.

12.2.4 Participatory evaluation methodologies and tools
Observations in variety trials commonly include several of the following:

Agronomic characteristics
o plant type: bush, climbing, tall, tillering, etc;
o earliness of flowering and plant maturity;
o resistance to pests, diseases;
o tolerance to drought and waterlogging, etc.
Yield characteristics
o total yield;
o yield stability (over sites, in time);
o yield of secondary products (straw, leaves, etc.).
Quality characteristics for consumption or other uses:
o colour of seed/skin and flesh;
o size of the product;
o processing quality and taste;
o storability of the seed.

The importance of these characteristics depends on the crop and the growing conditions. It is important to discuss with farmers the characteristic to be observed and the reason why. Farmers may come up with aspects that an outsider would not think of, such as leaf quality (for a vegetable) in sweet potato, the costs of certain cultivation methods such as the staking of climbing beans, and the food versus the market values of the produce. An outsider may, however, point out specific attributes such as differences in disease incidence. Group discussions with farmers are helpful to prioritize, add or delete characteristics.

Participatory variety evaluation can involve different methodologies and tools. Individual, independent selection may not be the best option for collecting information on variety adaptation, since not all farmers may express their observations in the same way. Joint evaluation involves the breeder/scientist visiting the farmer at different stages of crop development to discuss the performance of different varieties. Finally, group discussion is useful for the situations in which farmers' selection criteria are still unknown, to gain farmers' reactions to a relatively large number of alternatives, and overall interpretation of results. The technician facilitates the discussion by asking for opinions on different characteristics (Box 12.1).

There are three main tools to aid discussion of variety characteristics and participatory selection (see also Box 12.1).

Absolute evaluation: each variety is judged and assigned 'selected', 'rejected' or 'undecided' and the scores given by a large number of farmers can be averaged. For evaluating a relatively large number of materials in the field, farmers can be given for example, coloured ribbons or sticks to mark the materials they find attractive or would reject. The prominence of the markers may, however, result in a 'follow-my-leader' effect.

Matrix ranking: the alternatives are ranked from best-liked to least-liked. This is suitable for exploring the importance to the farmers of different variety characteristics or to compare variety behaviour over a number of characteristics.

Pair-wise ranking: each variety is judged better or worse compared to the local comparison variety or compared to all others in a set.

12.2.5 The need for statistical analysis

Most trials can be evaluated with these tools. In some cases, however, it may be useful to obtain statistical data, for example to calculate the response to better conditions (which is one of the yield stability parameters), or in order to obtain data that can be used to release the variety formally. Calculation of the statistical significance of difference is meant as a tool for situations in which the data do not provide obvious conclusions.

The need to include statistical designs has to be agreed between the farmers and scientist/technician. Complicated statistical designs are usually not required. In many situations, simple mean yields and deviations from the mean provide a good insight into the results. Replication to correct for a gradient in soil characteristics may be replaced by selecting a number of areas in the field of the local check variety for yield data (Figure 12.3).

Instead of experiments with replications on each farm, it is also possible to use the different farms in the experiments as replications, significantly reducing the complexity for each farmer. A ranking can be made of the results of the varieties on each of the farms (Figure 12.4). Such a ranking can be analysed statistically if necessary.

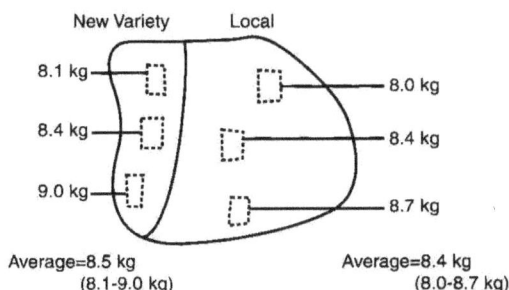

Figure 12.3 Sampling: in the case of a gradient in soil fertility or other forms of variation, the average of the harvest from a number of selected areas in the field will be a better estimate for the yield of the variety.

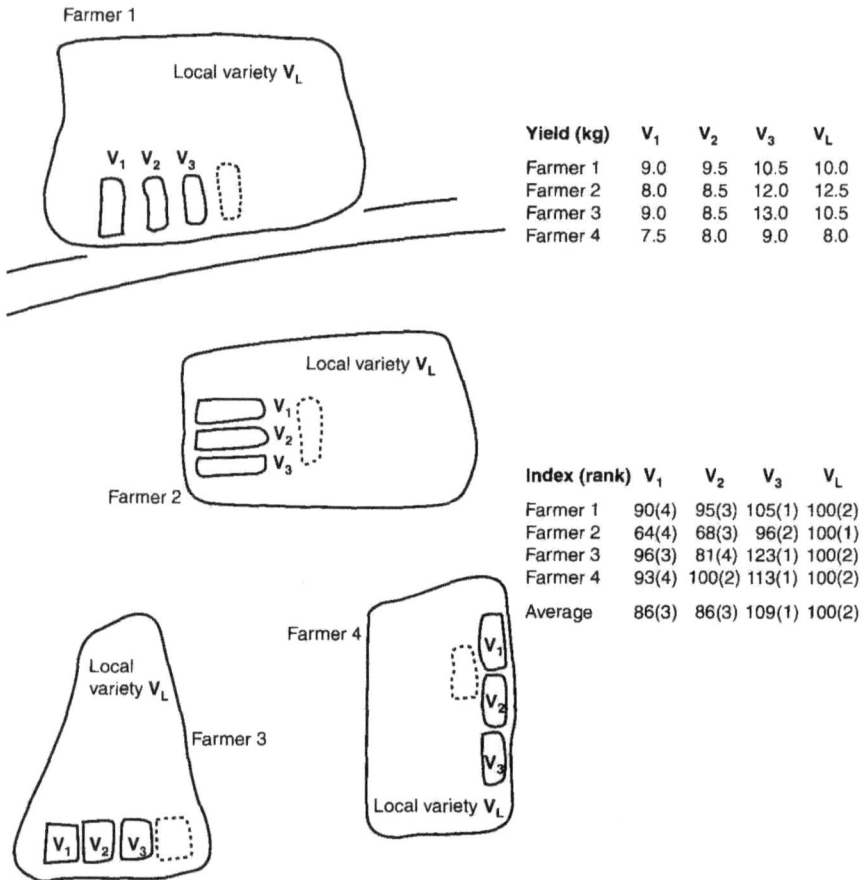

Yield (kg)	V_1	V_2	V_3	V_L
Farmer 1	9.0	9.5	10.5	10.0
Farmer 2	8.0	8.5	12.0	12.5
Farmer 3	9.0	8.5	13.0	10.5
Farmer 4	7.5	8.0	9.0	8.0

Index (rank)	V_1	V_2	V_3	V_L
Farmer 1	90(4)	95(3)	105(1)	100(2)
Farmer 2	64(4)	68(3)	96(2)	100(1)
Farmer 3	96(3)	81(4)	123(1)	100(2)
Farmer 4	93(4)	100(2)	113(1)	100(2)
Average	86(3)	86(3)	109(1)	100(2)

Figure 12.4 Testing three varieties, V1, V2 and V3 with four farmers by laying out small plots within the field planted with a local variety. A selected area of the local variety can be harvested for yield data of the local variety. Differences between farms are compensated for when all yields are indexed, assigning the value of 100 to the local variety.

Analysis becomes more complicated when, for example, plant densities differ within the trial due to roaming pigs (a random problem) or due to low seed quality of one of the varieties (a systematic problem).

12.2.6 Field layout and design

The choice of plot size, number of replications and locations depends on: the available seed; the farmers' time available; available land; soil variation at a particular location; and variation in environments that the germplasm is meant for. Within these limitations a balance should be struck between plot size, the number of varieties, number of replications and the number of sites.

Plot size
Preliminary observation trials may consist of many varieties in single-row plots or plot sizes of 2 – 5 square metres. Evaluation trials have to be larger for a good visual impact showing varietal differences. When the land is limited, it may be preferable to reduce the number of replications in favour of a larger plot size. If varieties on trial are very different in height, the plots need to have a border: the edges of the plots should not be affected by the neighbouring variety. Similarly with disease incidence: a diseased plot of a susceptible variety increases the pressure on the neighbouring varieties, and this is a particular problem when plots are small.

Sites
Ideally, the evaluation sites should represent the environmental variation encountered in the area for which the varieties are selected. This can involve differences in soil fertility, soil type, water availability (irrigation vs. rainfed), land preparation, etc. In practice, in participatory variety evaluation, the situation of the farms is largely determined by the collaborating farmers. However, if possible, it is worth while trying to include:

○ the extreme environments;
○ the 'average' environments; and
○ more intermediate environments.

If, for example, a range in water availability is expected, it may be possible to include (in order of importance): a rainfed field ('dry' extreme), fields with one or few irrigations (average) and repeated irrigation ('wet' extreme). Environments with different fertility levels are easy to simulate on one site with fertilizer. In such widely differing conditions one cannot use different sites as replications of the same trial. The different environments have to be considered as separate trials or treatments.

In on-farm trials it may be important to consider the field in which the farmer plants the variety. Farmers may plant the seed in marginal fields, when the evaluation is not their highest priority. On the other hand, it is very important that farmers do not treat the trial any better than their normal fields. It is common practice to try to have the 'best trial in the village', which means the cleanest and most vigorous crops. This kind of competition should be avoided, because some aspects of the varieties such as competitiveness against weeds may be hidden.

Replications
The number of replications is primarily determined by the soil variation at different locations in the same field. The more heterogeneous the soil, the more replications are required to obtain a reliable estimate of the 'true' average value. Two replications are commonly used as a minimum if this low number is balanced by more locations; three or four replications are

useful if the soil is very heterogeneous; and four or five replications are required if the goal is to determine seriously the agronomic value of a particular cultivar at a location.

Clearly, the workload increases with the number of replications, even if the total size of the trial remains the same.

12.2.7 Examples of planting designs

Example 1. Farmers can obtain small quantities (enough for 10m²) of seed of the varieties of their preference, selected from a demonstration trial. This allows for a first on-farm evaluation and for multiplication of the seed. When several farmers select the same varieties, an analysis of variance can be performed when farmers plant each seed variety in a 10m² plot (each farmer being a replication).

Example 2. A farmer has identified material in an on-farm evaluation. The initially small amount of seed has increased by harvesting the whole plot of the trial. The farmer wants to follow-up under 'real' conditions, i.e. he wants to plant part of his field with the new variety next to his old variety (Figure 12.3). Analysis is possible when, at harvesting, 1m² (or 5m²) plots are chosen at random and harvested separately. If this is done by several farmers, it can form the basis of analysis. It would even be better if the farmer planted the new variety in a strip through the centre of the field, with the old variety on either side.

Example 3. There are a large number of materials for demonstration purposes. An option is to plant plots of the local check variety at regular intervals between the varieties to be evaluated (see Figure 12.2) The yield of the check varieties alongside can be averaged and given a value of 100, the yield of other varieties can be related to those averages and thus be given a value of less or more than 100. Statistical procedures do exist for such designs, but the averages and indices provide a comprehensive overview of the results.

12.3 Case studies of variety improvement through selection

12.3.1 How to use the presented examples

A general introduction and experiences on participatory plant breeding and participatory variety selection have been presented in Chapters 2 and 8. This section presents a number of imaginary cases to illustrate how selection may be used to:

○ improve degenerated varieties;
○ respond to changing market needs;
○ respond to changing environmental conditions; and
○ increase yield in existing cropping systems.

In all these cases the existing landraces and varieties are taken as a starting point.

The case studies are definitely not meant to be 'rules-to-follow' but are suggestions for experimentation in comparable crops, in comparable situations. Selection within local varieties may have consequences for other characteristics: which are genetically related, but of which the genetic linkage may not be clear. An example of such an unwanted linkage is the one between male sterility in maize and susceptibility to Southern Maize Blight: selecting for male sterility eliminated any existing resistance to the fungus which caused a major disaster in the USA around 1970. Another common example is the linkage between yield and culinary quality (e.g. Green Revolution rice) and between yield of the main product and yield of secondary products, such as straw.

12.3.2 Case 1: improvement of degenerated varieties

A local variety is good and has been grown with success for many years. Through poor maintenance the variety has become too diverse: the variation in plant height among the plants increased after the introduction of some modern varieties of the crop in the area. The result is that the shorter plants do not develop sufficiently, that they get all kinds of fungus diseases and that the produce is of very poor quality.

Crops: cowpea (self-fertilized) and sesame (semi-cross-fertilized).

Strategy: regain the original local variety by performing a low selection pressure within the variety. Removing too tall and too short plants could be one method (*negative mass selection*), or selecting average size plants another (*positive mass selection*). Inbreeding depression is not a great risk in these crops, but the original level of diversity should be regained, so selecting a sufficiently large number of plants is essential.

Where the original variety was very diverse in plant height even before the introduction of modern varieties, negative mass selection can work well. This removes the extreme tall and small types and will maintain the original diversity. Where the original variety was fairly uniform in plant height and the variation in the degenerated variety not excessive, positive mass selection may be most effective. Selection of a few plants which have similar plant height will provide the basis for further multiplication.

A refinement may be necessary when the degenerated variety is very heterogeneous: *ear-to-row selection* (see Chapter 8). Good plants, i.e. those that have the correct plant height and all other important characteristics, should be harvested separately. The seeds from each selected plant are planted in rows in a separate selection field. The rows that show all important characteristics quite uniformly should be harvested and bulked. Because not all characteristics may be visible during that particular season, at least 50 lines should be selected, which will jointly form the 'improved local

203

variety'. When the sesame lines show considerable diversity within the lines (i.e within the rows with plants derived from one ear), this means that there are genetically heterogeneous plants in the population, which indicates that the natural crossing percentage is quite high. In this case, the method to make the variety more uniform for its essential agronomic characteristics is to place paper bags over (part) of the selected plant before flowering, sowing the seeds of each ear in separate rows and combining the seed harvested from the similar and uniform lines only, possibly using the *remnant seed method* (Chapter 8).

12.3.3 Case 2: Generating uniformity for the market
A crop has been grown with success for many years. The produce is partly home consumed, partly marketed. With cities expanding, the market is becoming more important and the demand has changed from mixed produce, to uniform-looking grains for processing or direct consumption. The local varieties are good, except for the variation in seed colour.

Crops: beans (self-fertilized) and maize (cross-fertilized)

Strategy: Selecting within the local varieties should do the job. This is easy in beans, because they are principally a self-fertilizing crop with mostly genetically homogeneous plants: the target seed colour (e.g. white seeds) can be picked and multiplied. This could result in a uniform seed colour straight away, but it may be necessary to have one cycle of line selection, i.e. planting the seeds of each selected plant in a row, to check on possible segregation. If the target seed colour is already an important constituent of the existing mixtures, this method may suffice because there is likely to be a sufficient variation within white coloured beans with respect to important agronomic characteristics. When white seeds are rare in the local bean variety, selecting the white seeds may result only in a genetically very narrow (for colour and other characteristics) white-seeded selection. Some varieties owe their yield stability to their genetic variation. If this is the case, and uniform-looking produce is an important market requirement, other white-seeded types from landraces within the region or from research stations could be tested for their ability to fit in with the local variety.

When a local variety of a cross-fertilizer like maize consists of different seed colours, e.g. white, yellow, red and purple, different strategies could be followed. A primary concern is to isolate the selection field from any seed field where coloured seeds appear. When the target colour (e.g. white) is dominant, *mass selection* is effective: planting only white seeds for a number of seasons will yield a stable white variety. Some purple-coloured seeds will continue to appear, but in ever-declining numbers when cobs with purple grains are removed. Alternatively, one cycle of *selfing with remnant seed method* would do the job. This means that the young cobs of a large number (at least 500) of plants are bagged before the silking. When the silk is ripe the tassel of the same plant is shaken into the bag (selfing) (see Figure 12.5). These cobs are selected at harvest for uniform seed colour (which is a characteristic of the mother plant). The seed of these cobs (at least 400) are bulked and planted in a recombination block, i.e. they are planted together to allow free pollination and fertilization between them. Since these plants were products of selfing, weak plants may appear, but this does not require attention. The best procedure is to detassel about 75 per cent of these plants at random in the recombination block (i.e. remove the male flower) and harvest from those plants only the uniformly coloured seed. These detasselled plants have been cross-pollinated with the 25 per cent of the plants which were not detasselled. This procedure should eliminate inbreeding.

12.3.4 Case 3: Responding to changing environmental conditions

A local variety performs well, but changing agroecological conditions make it less suitable for further cultivation. Irrigation infrastructure has resulted in year-round production and disease levels have increased in all crops. The arrival of a new disease causes the local varieties to perform unsatisfactorily. Resistance against the disease is not found in the local varieties.

Crops: wheat (self-fertilizing) and pearl millet (semi-cross-fertilized).

Characteristic: resistance to smut.

Strategy: Compared to the previous examples, this problem is much more difficult to solve. In the former examples simple selection methods were used on good local varieties in which the desirable genes were present. Here, new characteristics have to be introduced into the crop; new genes need to be incorporated.

Option 1. Participatory Variety Selection: check whether resistant varieties exist which (a) perform well in the region, (b) fit in with the cropping system and (c) produce grain and byproducts (e.g. straw) acceptable to the local people. The national agricultural research system may be able to trace varieties that can be used. Local varieties from other regions of the country, which may have been developed to resist the disease already, should not be overlooked.

From wheat varieties that perform well, seed can be harvested for further testing and multiplication, since it is a self-fertilizer. Pearl millet seed from an evaluation trial will have been fertilized by all the varieties around it and harvesting this seed would produce a very diverse population unlikely to retain the important characteristics for which the variety has been selected. This means that seed of the selected variety would have to be obtained again, or that seed would need to be retained before the screening trial (see Case 1). This PVS option is the quickest way to solve an acute problem, but may not yield a new variety that is perfectly adapted. If the

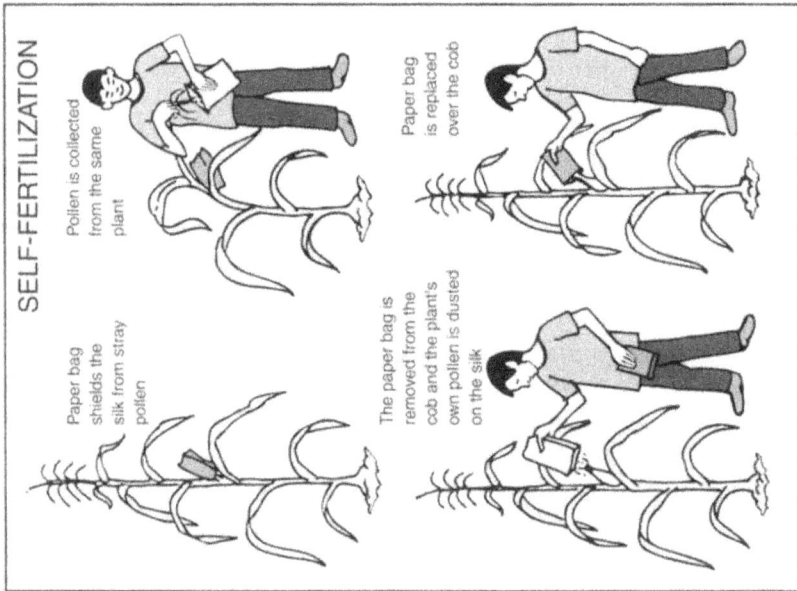

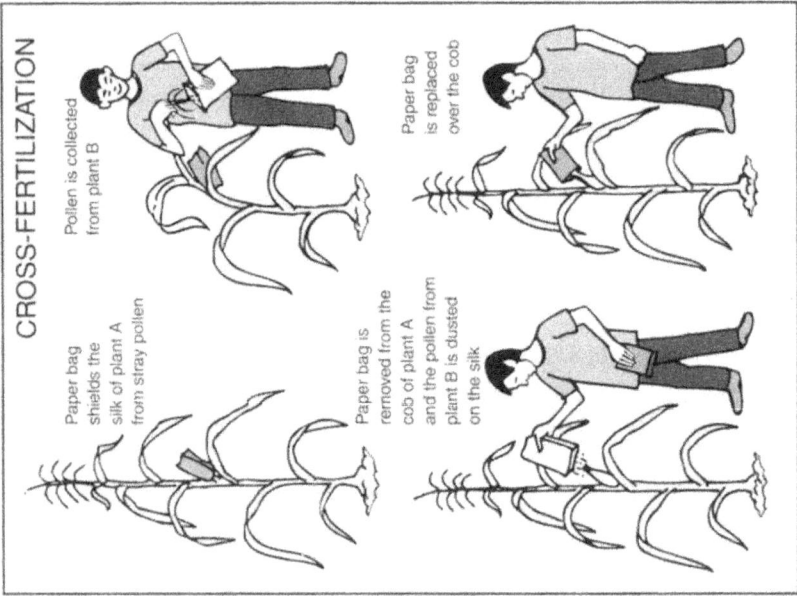

Figure 12.5 Selfing and crossing in maize.

problem is severe, such a sub-optimal solution will have to be accepted while more long-lasting solutions are being tried out.

Option 2. Participatory Plant Breeding: There may be a source of resistance in non-adapted varieties. In such a case, breeding is the only solution, i.e. crossing the source of resistance with the local variety. At this point there is a significant difference between self-fertilizing crops (wheat) and cross-fertilizers (pearl millet).

In pearl millet, the source of resistance may be planted in rows within the local variety. The two will cross naturally, and selection for the important characteristics for cultivation and use could be done when the crossing population is subjected to the disease next season. Plants without the resistant genes will become diseased, and if the farmer carefully selects seeds from healthy plants only, the number of plants with the resistant genes will increase rapidly over a number of seasons. Where available, different sources of resistance against the disease should be used at the same time, since such different sources may diversify the resistance genetically, and possibly make the resulting resistance more long-lasting. This method can be used with fairly diverse local varieties. When the local variety is quite uniform, this method may result in an unacceptable variation, and real plant breeding procedures may have to be followed.

With self-fertilizing crops, crossing is a specialized job, the details of which go beyond the scope of this book. The following stages have to be followed:

(1) hand pollination of a number of plants;

(2) selection during the first generation, whereby a number of characteristics cannot be observed easily because of hybrid vigour; and

(3) more strict selection of the stabilized genotypes after four or five seasons. Depending on the interest, commitment and knowledge of the community, all steps could be done by them. In other cases, the crossing could be done by a national research programme. If the national programme can raise the first generations in the target region, farmers could assist in the selection of step (2). Otherwise, the national programme should multiply the crossed seed through a so-called *single seed descent* method, whereby selection is avoided until stabilized genotypes are obtained. Such a population of plants (after five or six seasons) could be sent to the region for farmer-selection. The selection should be done under high disease pressure.

Breeders are used to making crosses between two varieties. When in this case the original variety is a genetically diverse landrace, the crosses have to be made between a large number of plants from that landrace, and the source of resistance (preferably, multiple sources). Breeders may be hesitant to take up such a task.

207

12.3.5 Case 4: Increasing yield in existing cropping systems

Local varieties may do well, but lag behind in producing enough food for a growing population: hence increased yields are needed.

Crops: beans (self-fertilized) and maize (cross-fertilized).

Strategy: Unlike the earlier examples with clearly defined targets, this need is a continuous one. Secondly, the problem is rather poorly defined from a plant breeder's point of view: yield is a complex of a large number of factors. Compared to seed colour, which is easy to observe and genetically well defined in one or a few genes (high heritability), the yield of a plant is strongly influenced by the environment and by a wide range of genes. In breeders' terms: yield has a low heritability, i.e. when high-yielding plants are selected, this does not mean that the same high yield will be observed in the offspring. Strategies are numerous:

○ Different selection methods (Chapter 8) within a variety may very well increase yields, but there are two major setbacks: these methods will quickly reach a ceiling beyond which further selection does not help any more. There have been studies showing that this ceiling may be around a 15 per cent yield increase. Secondly, selection within a population necessarily means reducing the genetic variation, and may affect yield stability. In cross-fertilized crops it may easily lead to inbreeding. Negative and positive mass selection are, however, relatively easy to carry out with farmers, especially with self-fertilizing crops (which do not show an inbreeding effect) and the effort may pay off with higher yields.

○ PVS: get new materials from outside, test them under local conditions and select the best for introduction in the community.

○ PVS-PPB: get new varieties from outside, test them under local conditions and mix them in with the local varieties. This will enrich the diversity, which will allow the farmers to select from a wider population and thus mould the variety better to their needs. This could result in the gradual removal of other types from the mixtures (beans), or the gradual shift of the landrace in a certain direction. This method could also adapt the variety to gradual increases in the use of inputs such as fertilizers. This method is not strictly PVS because it does not test new varieties, and it is not PPB because there is no (intentional) crossing followed by selection.

○ PPB: when the characters that are responsible for the limitations in yield are known (e.g. drought tolerance, disease resistance, competition against weeds), parents that excel in coping with these limitations can be sought for crossing with the local variety and selecting a new variety. When yield as such is the selection criterion, statistical design and analysis of trials can hardly ever be avoided (see Section 12.2.5). There may be other characteristics, i.e. more heritable ones, that are linked with yield, and are easier to select for. For example, in maize it is known that plants with increased time between the male flowering and the tasselling of the cob when under stress are less drought tolerant. Such secondary characteristics can be used as selection criteria for some time, until a ceiling of possible yield increase is reached. When breeding continues, however, and advances tend to slow down, formal methods have to be used to introduce new genetic variation which may create a higher ceiling which can be reached after selection.

A common aspect of all these strategies is that if farmers do not want to become entirely dependent on formal plant breeding, they should obtain the skills and the interest in experimenting independently from the researchers.

12.4 Experimentation with seed production

Experimentation with seed production usually aims at improving the quality and/or quantity of the locally produced seeds. To measure the impact of the improved treatments, seed quality evaluation is required (Chapter 10). Improved agronomic practices in the field, and better control of pests and diseases can be evaluated for their effect on the quality and quantity of the seed produced. These practices could include fertilizer, pest control, spacing and planting date trials. Seed quality characteristics could also be measured, through trials to compare drying and storage treatments (such as local insect repellents and insecticides). Standard farmers' practices can be compared with alternatives that may come up during group discussions (Box 12.2). Always use one seed lot for such experiments, or several well defined ones, all getting all the treatments.

A more complicated but very interesting exercise is the comparison of different seed selection techniques. The effects of randomly taking seeds from a harvest, selecting heads in the field and selections from an ear-to-row maintenance field are useful to determine their effects on varietal identity and uniformity, and yield. Such tests, however, need field trials as described in Section 12.2.

Another reason to experiment with seed production practices is to evaluate the effectiveness of local methods. Farmers in Eritrea, for example, wash their sorghum seeds with cattle urine to reduce seed transmission of smut. Local methods of insect control, such as mixing beans with vegetable oil or ash, covering stored potatoes with herbs etc., should be tested alongside 'modern' insecticides and untreated seeds in experiments. Apart from providing a useful evaluation of these methods, local experimentation also increases the commitment of the collaborators.

12.4.1 Case 5: Improving farm-produced bean seed in Costa Rica

Assessment. In a discussion, farmers stated that their farm-saved seed from the local bean variety Parriteño gave very unsatisfactory results due to irregularity in the plant stand (Cárdenas and Alvarez, 1996). Seeds purchased from the Ministry of Agriculture produce a more regular crop. However, this purchased variety is not liked for home consumption and does not fetch a good price on the market. Bean seeds are stored by the farmer for over six months in jute bags. The climate during storage is warm and rainy.

Problem identification. The farmers decided to analyse seed quality in a group experiment. Each farmer brought 200 of his or her seeds to germinate in the horticultural nursery of the school. Since the test was carried out in the dry season, the school children took care of the nursery. Small plots of 2m² were prepared for each sample. Seed from the Ministry of Agriculture was obtained and sown every fourth plot for comparison. One week later, farmers gathered and checked seed germination and emergence. Digging up the non-germinated seeds showed that these seeds had produced mouldy sprouts or had not germinated at all.

The findings were analysed with the farmers, and in the discussions problems with fungal diseases in the field and insects in storage became apparent. Farmers were interested in exploring ways to produce healthier seed.

Planning. Farmers decided to arrange for a seed production plot which combined the purpose of experimentation and seed production. Considering the costs of the experiment (chemicals and fertilizer), they decided on a common seed production plot. They used seed from the farmers with the best germination results in the nursery test; these farmers would be given back double the quantity of seeds after harvesting the trial. The rest of the seed would be divided among all farmers. The community plot was installed in the school garden.

Implementation. Farmers decided to take the following measures:

○ The distance between the plants was twice the normal planting distance.
○ During the growing season, the crop was closely observed. When the plants were fully developed (but the canopy not completely closed), diseased plants were removed.
○ When the first signs of a fungus attack appeared, plants were sprayed with a fungicide.
○ At harvest, the plants were dried on a concrete floor and taken inside during occasional rains, not left on stacks in the field, as is common in the area.
○ Once seeds were well dried, they were stored in barrels, and an insecticide was added before the barrels were hermetically closed and placed in a cool place in the house of one of the farmers.
○ Before planting, seeds were carefully selected; only well-formed, undamaged seeds of similar size were used for planting; shrivelled seeds were removed.

Evaluation. A germination test was again carried out in the school garden nursery with three samples of 200 seeds from the experimental seed lot to compare with seed obtained from the outlet selling the material from the Ministry of Agriculture. The germination data from the local varieties had improved considerably and few mouldy non-germinated seeds were found. It could not be concluded whether it was the precautions in the field or the better storage which was the reason for the better performance of the seed.

12.4.2 Case 6: Improving on-farm seed production and the use of millets

An NGO operating in West Africa collected some samples of millet during a diagnostic survey in a 300–400mm rainfall area (Osborne, 1990). During the survey it was observed that, unlike the modern varieties, the traditional millet varieties were well adapted to the length of the growing season and the culinary preferences of the farmers, but that plant stands in the field were very variable. The collected samples were analysed. The fraction of small and damaged seeds was considerable. After removal of this portion, two samples of 200 seeds from each collected seed lot were germinated: one sample with a so-called rolled-towel test (see Chapter 10) and another sample in a calabash with moist sand.

Results of the analysis were discussed with the farmers from whom the seed had been collected.

Farmers showed interest in carrying out the simple seed testing in a calabash filled with moist sand. Individual farmers and NGO technicians also decided to mark out a 20 × 20m seed plot in the centre of their field where they thinned the millet to one plant per hill and removed diseased plants. At harvest they selected plants for seed based on colour and vigour, absence of disease, high tillering, and having many medium to large heads. Utilizing the harvested seed and the sieving, farmers achieved 30 per cent yield increases in the following year. Other farmers were informed through group meetings where the individual farmers were invited to relate their experiences. In 1992 as many as 50 farmers in five villages produced 850kg of seed.

12.4.3 Case 7: Building on traditional seed flows

Peru has three distinct potato-growing zones: the highland (3500-4200m above sea level), the inter-Andean valley bottoms (2500-3500m) and the desert coast. Commercial large-scale potato production with improved varieties takes place principally in the last two zones. The potato production in the highlands is a small-scale farmer activity. These farmers form the majority of the Peruvian potato producers.

In a special project within the Peruvian National Potato Programme in 1983-88 it was concluded from earlier research and surveys that the small-scale farmers grow a wide range of varieties, often in mixtures, including improved varieties and native varieties long discarded by the big farmers (Scheidegger et al., 1989). The planting of mixtures and the saving of one's own seed seem to be important elements of a risk-avoidance strategy and provide a more varied diet. Research had shown that these farmers could improve their yields with better-quality seed, but in general the farmers' own seed was of such good quality that a high price for new seed was not justified. In order to match the farmers' conditions, the project chose to work with a wide variety of older improved varieties.

Surveys showed that farmers generally renew their seed by exchanging or purchasing small quantities of seed locally, or from merchants, and sometimes from commercial seed growers, but that certified seed was not a relevant source of material. For this reason the programme started with the sale of 20kg bags of seed directly to these small farmers, via weekly local fairs. Earlier surveys had also identified that flow of seed moved from upland zones to lower areas where it is more difficult to produce and store good-quality seed. Farmers from the lower zones need to renew their seed every two to three years.

This provided the opportunity for the programme to establish seed plots with eight individual farmers and four communities in these key upland areas in 1985/86. The follow-up on this activity showed that distribution from these key farmers to other communities and farmers was quite successful. In 1986/87, the communities had sold seed to 11 other communities and 74 individual farmers. The eight farmers had sold seed to 26 other individual farmers. This showed that building on the local system can be quite effective if key-farmers can be identified. Rural development projects and NGOs can play a role in scaling up such activities. Unfortunately this programme had to be discontinued because of terrorism in the country at that time.

12.5 Experimentation with seed quality measurements

12.5.1 Standardization

When problems occur in crop production that might be due to the quality of the seed it may be useful to test the seed quality objectively. Alternatively, it may be necessary to compare the seed quality in different treatments of an experiment. Seed quality aspects may also need to be measured in order to evaluate a seed assistance programme.

Seed testing can be done at various levels of precision. One always has to balance the need for tests to be reproducible with their being representative of farming conditions (Chapter 10). Official seed tests are designed to be highly reproducible. Whether a test is performed in one or other of the ISTA-accredited laboratories should not make any difference to the results. This is important because countries may base an import licence on such reports. Viability measured in a seed laboratory may significantly deviate, however, from the emergence in a farmer's field. Viability (commonly called germination) percentages thus do not present the full picture, since farmers are more interested in field emergence and seedling vigour.

It remains important, however, to standardize conditions to a certain level, and seed testing experiments are performed in many projects. This is illustrated below on germination tests.

12.5.2 Seed testing to evaluate observed problems

Example: field emergence is very low in some farms, and not in others. It might well be that the viability of the seed is at stake. Alternatively, the farmers used different planting techniques (planting depth), or one field might have been infested with cutworms or Fusarium fungi that killed the seedlings. It is necessary to compare the seed used by different farmers around planting time. If farmers still have some left-over seed this can be tested and the only requirement is that the seed samples should be tested under the same conditions. Whether tested in an official seed laboratory, or under the conditions of any of the farmers is not fundamentally important: if the viability is the same, the results will be the same. It is, however, important not to make the conditions too hard for the seeds because if seeds do not emerge in either seed lot no result is obtained.

The best local solution may be to put a known number of seeds at a uniform depth in well-mixed sand or light soil in a container or a nursery bed. The bed should be covered to avoid direct sunlight, and regularly watered to give all seeds the opportunity to germinate. Less optimal conditions may do, as long as all seeds of the seed lots are treated in the same way.

12.5.3 Seed testing as part of an experiment

When a seed drying or storage experiment has to be evaluated, it may suffice to test the seeds of the different treatments at some point in time. Again, a fairly rough trial may suffice. It may well be that different farmers are involved in the trial and that germination of different treatments has to be measured by different collaborators at the same time. Since ambient temperature will be roughly the same, local germination tests may be carried out. In the analysis the absolute figures may not tally due to the different methods, but the ranking of the results obtained by each farmer will probably be the same.

Quite often, however, it is more interesting to know the decline in viability during storage, or to know the effect of the initial viability, or the variety, on this decline. This means that absolute data have to be compared of tests that are carried out at different moments, in different seasons and temperatures. For this purpose, germination conditions have to be standardized to a large extent.

As a general rule: the more experiments have to be compared in time and space, the more testing should be standardized.

12.5.4 Seed testing as an evaluation tool

It may be clear from the above that when seed quality is considered a tool for evaluating the effectiveness of a project, seed tests have to be standardized. Tests have to be compared over an extended period of time. Secondly, local testing procedures may be refined by the project (e.g. use of different soil, more uniform conditions), so that test results from the start and the end of a project cannot be compared at all. In such cases it is wise to contact an established seed laboratory to provide you with data.

12.5.5 Testing test methods

It may be very illustrative in the initial stages to perform a test on different viability tests. One seed lot may be well mixed, and divided and sown in a number of different conditions, ranging from a roughly ploughed field to a rolled towel test inside a house. Intermediate conditions may be sheltered/open nurseries used for germinating vegetable seedlings, and containers. Comparison of the treatments may yield both a seed quality awareness and an experimenting attitude with the co-operators and a good local germination technique.

PART D

CROP-SPECIFIC OPTIONS

Barley

Other names	orge, cebada, cevada	Market potential	poor
Propagation	self-fertilizing	Ease of seed	
Storage	good	production	easy

General description

Barley (*Hordeum vulgare* L.) is essentially an annual grass adapted to an arid climate with a rainfall season of limited duration, and was domesticated in the Near East. Its drought tolerance is greater than that of durum wheat. For that reason, it is grown in West Asia and North Africa in regions with annual rainfall of less than 400mm, where it is used for human consumption (flat breads) and animal feed. In temperate regions, barley is grown for malt for beer brewing. Its fodder is a valuable byproduct. Barley also has a good tolerance to cold, heat, and salinity, which makes it a suitable crop for adverse environments.

Barley is an erect plant with a height of up to 1.5m in the case of landraces, and a height of 1m or less in the case of modern varieties. It tillers profusely if plant density is low and growing conditions are favourable. A terminal ear of 7.5-10cm length contains two or six rows of kernels.

Agronomic requirements

Barley is mainly grown in temperate climates; production in lowland tropical environments is not possible. In mountainous tropical areas or in countries with a winter season (Bangladesh, India) barley can be grown in environments unsuitable for wheat production.

Rusts, mildew (*Erysiphe graminis*), covered smut (*Ustilago hordei*) and foot rot (*Helminthosporium sativum*) are the most important diseases. A seed rate of 60-100kg/ha is recommended, with a spacing of 20cm between rows. Seeds are sown 2-4cm deep. Landraces, which are usually tall and have a weak stem, lodge easily, but since the straw is also an important product, farmers may still prefer these types to shorter modern varieties.

Storabilty of the seed is poor, especially during the humid wet summer seasons, when special precautions have to be taken.

Seed production

Selection

Seed production is best done on heavier soils with good water-holding capacity and good surface drainage. Contamination with weeds or other crops should be avoided. Fields that have been used in the previous season for other small grains should not be used, to avoid seed contamination and diseases. Sow in sufficiently high density to suppress growth of emerging weeds.

Barley is basically a self-fertilized plant, but a significant degree of cross-fertilization may occur, especially under stress conditions (e.g. drought). Barley varieties are therefore much less homogeneous than wheat or rice varieties. This may be an advantage under marginal conditions. Assisting farmers in variety development through pure line selection should thus be done with care.

Because of significant amounts of cross-fertilization, greater distance between seed fields may be advisable compared to other self-fertilized small grains.

Seed-transmitted diseases
The most common seed-transmitted diseases are covered smut (*Ustilago hordei*) and stripe disease (*Pyrenophora graminea*, syn. *Helminthosporium gramineum*). Scalp (*Rhynchosporium secalis*) is transmitted by seed, as well as by plant debris.

Harvesting and cleaning
Threshing barley is a relatively easy operation: the kernels are released completely from the ears. Winnowing will remove most of the unwanted plant debris. Wild oats (*Avena fatua* and related species) and admixtures of other cereals are difficult to remove. Processing the seed with an air-screen cleaner is highly recommended: it is one of the most effective tools for reducing the field (see under 'wheat'). The storage behaviour of barley is similar to wheat, but where this crop is generally grown in more marginal (drier) areas, problems with germination are encountered less often. This may, however, not be true for high altitude cultivation in equatorial areas.

Options for improvement
Attempts to improve barley seed systems may concentrate on the varieties. Since most barley is produced under marginal conditions, formal breeding programmes may encounter difficulties in selecting better varieties for the wide range of specific conditions. Also, the different uses of the crop (grain, straw, stubble) may pose particular problems for breeders. Participatory Plant Breeding may thus be a suitable approach. The International Centre for Agricultural Research in the Dry Areas (ICARDA) in Syria has a useful collection of germplasm.

Because barley is often considered a 'poor man's crop', commercialization of seed supply is very difficult compared to wheat. Especially under dry conditions, farmers are technically very capable of producing their own seed, but they may face problems with seed security. In high-altitude tropical areas barley is often produced for the beer industry, and there is then a commercial demand for seed, which is commonly catered for by the industry itself.

Bean

Other names	Common bean, kidney bean, French bean, haricot, *frijol*	Market potential	poor
		Ease of seed production	quite easy
Propagation	self-fertilized		
Storage	good		

General description

Beans (*Phaseolus vulgaris*) are the most important food legume in tropical Africa. They may be erect, semi-erect, trailing or a climbing legume. Bush types reach heights of 30–60cm. Phaseolus bean is a self-fertilized plant. Beans are grown for pods (snap beans) or for their immature (green shell) or dry seeds. Varieties differ greatly in seed colour and seed size. These characteristics are decisive for the market potential of the product. Even where varietal mixtures are preferred in growing the crop, uniform-coloured beans may fetch a higher price.

Agronomic requirements

High temperatures cause dropping of flower buds, and really humid climates favour the many diseases of this crop, thus making the medium altitude areas (800–2100m) most appropriate for cultivation. Bush types are preferred in large-scale and mechanized planting, whereas climbing types are most appropriate for intensive farming, especially on slopes.

Beans are often inter-cropped with maize, sorghum or root crops. Beans require a well-drained soil and sufficient phosphorus and potassium. Even though root nodules supply nitrogen, some initial nitrogen fertilizer may considerably improve yield.

Bush varieties are planted in rows 50–60cm apart, with a spacing of 10–15cm within the row, resulting in 100 000–150 000 plants per ha. Large seeded varieties thus require 100kg of seeds per ha, small-seeded approximately 60kg. Beans are more vulnerable to weed competition than quick-growing cowpea and soybean. Weeds may easily reduce bean yields to a bare minimum. Pods are ripe when they lose their green colour. Plants are normally lifted and often left in the field or taken to the homestead for after-ripening.

A wide range of diseases attack beans. Both bacteria (common bacterial blight, halo blight), fungi (anthracnose, angular leaf spot, rust), and viruses (common bean mosaic virus) may cause serious crop losses. Most of these diseases are seed transmitted. Of the field insects, bean fly can be a major threat to bean production.

Seed production

Selection

Being a self-fertilizing crop, variety maintenance selection is easy. Off-types are easily observed by plant type, flower colour and maturity period. After harvesting, seed size, shape and colour are clear indicators of off-types. When a mixture has to be purified, a simple line selection will be very effective.

Seed-transmitted diseases

o Common bacterial blight (*Xanthomonas campestris*): water-soaked brown spots with bright yellow borders, extending to large dead areas with yellow borders. Infected seeds are sometimes wrinkled and the 'eye' may be discoloured. May infect beans through seed, soil and water splashes.

o Halo blight (*Pseudomonas syringae*) occurs in cooler areas compared to common bacterial blight. Symptoms are diverse, starting with water-soaked spots on leaves, stems and pods. Leaves may get virus-like symptoms, and the most typical symptom is the 'grease' spots. Crop rotation and seed selection are the main control measures. Chemical treatment (antibiotics) is not recommended.

o Bean common mosaic virus (BCMV) can be a very important seed-transmitted disease. It is characterized by a vein-banding mosaic, sometimes accompanied by leaf malformation. Resistance is the most important control measure. Alternatively, non-continuous cropping is effective to reduce damage by BCMV, i.e. having a period in the year without hosts of the virus (apart from beans, mungbean and several other wild and cultivated legumes).

o Angular leaf spot (*Phaeoisariopsis griseola*) causes reddish spots with typical angular shapes. In disease-infested areas it is not necessary to treat the seeds, because it also spreads effectively from plant debris by wind and rain splashes.

o Cool wet weather favours the development of elongated dark-red cankers in the stem and the leaf veins, caused by anthracnose (*Colletotrichum lindemuthianum*), which can be very destructive. Planning the ripening of the crop in dry conditions is the best way to reduce this disease, apart from Thiram seed treatment.

Measures. Because of the importance of diseases in beans, some seed production methods have to be taken seriously:

o Planting early, i.e. just before the bulk crops are planted in the area, reduces the incidence of air-borne diseases. This may contradict the requirement that planting should be planned in such a way as to harvest in a dry season.

220

o To avoid soil-borne diseases such as common bacterial blight and diseases that remain in the stubble (especially fungi), crop rotation is extremely important.

o In order to avoid the spread of diseases through drops of water, it is better to widen the row-spacing (e.g. to 80cm) and reduce the spacing within the row to get the required plant population. Regular roguing of diseased plants during the vegetative stage is necessary, but should never be done early in the morning or at other times when the foliage is wet.

o The common practice of using the early pods on a plant for fresh beans or green seeds, leaving the later-developing pods for seed has to be strongly advised against. The last pods have the highest incidence of seed-transmitted diseases.

o Chemical treatment of seeds against fungi (e.g. thiram, Vitavax) is highly advisable; blanket treatment with antibiotics against the important bacterial diseases is, however, not recommended.

Where possible, bean seed should be produced in isolated areas or off season (in most cases under irrigation). In bean fly-infested areas on the other hand it is not advised to have bean crops in the off-season. The longer the season without a host the better it is. Destruction of the stubble and seed treatment (endosulfan) can significantly reduce pest incidence.

Storage
A wide range of bruchids are responsible for damaging stored beans. Storage in the pods, mixing threshed seeds with sand, ash, rice husk, and vegetable oil (5–10ml/kg), or in airtight containers are all effective to some extent. Actellic is a cheap and relatively safe insecticide that will combat bruchids and most other storage insects. When dried properly, bean seeds store quite well.

Options for improvement
Beans are bulky seeds and relatively easy to produce by experienced bean farmers. Commercialization of bean seed production is only viable where seed-borne diseases or bean fly are rampant, and where these can be kept at bay by specialized seed producers through field control and seed treatment. In all other cases, seed quality is best improved by general extension messages to all bean farmers, based on selection and identification of diseases.

Cassava

Other names	Manioc, tapioca, *yuca*, *mandioca*, *karapendalamu, mhogo*	Market potential seed Ease of seed production	poor easy
Propagation	vegetative		
Storage seed	fair		

General description

Cassava (*Manihot esculentum*) is a perennial shrub of 1.5–4m high. It forms a tap root and secondary roots that form the storage organs, and these vary in size and shape considerably.

The growth of the main stem stops 2–5 months after planting and inflorescences are formed. Starch accumulation starts around one month after planting. The male and female flowers in the same inflorescence do not open at the same time, but self-fertilization occurs between different inforescences.

Agronomy

Cassava is known for its drought tolerance and good yielding capacity in poor soils. The crop is grown in warm tropical environments with rainfall from 500 to 3000mm and more. When drought occurs, the plant may form smaller leaves and drop older ones to reduce transpiration. When water is available again, normal leaves are formed afresh. Cassava takes up large quantities of soil nutrients, which can lead to soil erosion when these are not properly compensated. Although growing in high rainfall areas, cassava needs well-drained soils to avoid rotting of the roots.

Seed production

Selection

Cassava is propagated by stem cuttings. Propagation through seed results in heterogeneous offspring, and is done in breeding programmes only. The vegetative propagation results in identical offspring in which selection for genetic improvement is not effective. Cassava germplasm for local testing is available from CIAT and IITA.

Multiplication

Cuttings are preferably taken from the middle parts of the stem; they should be well lignified and 4 or 5 nodes (20–30cm) long. One-year-old stems can yield 1–5 cuttings under unfavourable, and 10–30 cuttings under favourable conditions. The cuttings can be stored for a few days; whole stems for up to three months in a cool and shady place. They can be planted upright, slanted or horizontal just under the soil surface in moist soils, and deeper in dry conditions. The planting density in monocrops is

10–15 000 plants/ha. Growing 50 000 plants/ha yields more cuttings as under these conditions more stems (and fewer marketable roots) are produced. Also, reducing the size of the cutting to three nodes increases the multiplication.

Pests and diseases
Maintaining the planting material free from systemic diseases is, like other vegetatively propagated crops, a major concern. Major cassava diseases are African mosaic virus (AMC), Cassava bacteriosis (CBB) and leaf spot (*Cercospora sp.*). Larvae of the fruitfly (*Anastrepha sp.*) eat their way through the stem. A secondary bacterial infection with *Erwinia sp.* makes the stems unsuitable for propagation. In some countries spider mites are a major pest that can also be transmitted by cuttings.

Options for improvement
The quality of cassava planting material is commonly a problem only with respect to virus infections. Selecting 'clean' plants or 'clean' production areas may be a useful exercise, especially when performed in a larger area, because otherwise re-infection may quickly occur. Furthermore, by selecting varieties with a short 'neck' (the root between the plant and the storage root), appropriate processing qualities and disease and pest resistance can improve cassava production.

Chickpea

Other names	Gram, *pois-chiche*, *garbanzo*, hummus (Arabic), *chana* (Hindi)	Market potential seed	poor
		Ease of seed production	easy
Propagation	self-fertilizing		
Storage seed	good		

General description
Chickpea (*Cicer arietinum*) is a very drought-resistant crop. It requires light soils and is commonly grown in the winter season in sub-tropical regions in Asia and Africa. It is an erect or spreading bush of 25–60cm height. The seeds are used in curries; the flour has many specialized uses. Chickpea dishes (homs) are an important source of protein. Chickpea is a strict self-fertilizing plant, i.e. cross pollination which is still common in many other self-fertilizers is very uncommon.

Agronomic requirements
Being drought tolerant, chickpea is normally grown under semi-arid conditions. Too much moisture may even reduce yields by promoting excessive vegetative growth. The maturity period of 4–6 months (depending on the

variety) normally secures harvesting under dry conditions. Planting in rows 40cm apart and 15cm spacing within the row results in a plant population of 170 000, using 40–60kg of seed/ha. Chickpea is an effective nodulator provided that the appropriate rhizobium is available. Since chickpea is the least sensitive to shattering of the legume crops, delaying harvesting is possible until the crop is fully dried.

Seed production

Selection
Because of the crop's strict self-fertilizing breeding, varietal admixtures occur only as a result of mechanical admixture. Purifying such a mixture is very easy. Major selection criteria are plant habit (erect, spreading), and seed characteristics (colour and size).

Seed-transmitted diseases
o Chickpea wilt (Fusarium species) causes rapid drooping of entire plants, and is both soil and seed borne. It is very difficult to eradicate once introduced in an area. A check on seed-borne Fusarium is therefore important when introducing the crop in a new area.
o *Ascochyta* blight is a common disease in most chickpea-growing areas, but not in Southern Africa. It causes brown elongated spots with dark brown patches on stems and leaves.

Storage
Chickpea seeds store well when dried to at least 10 per cent moisture content. Bruchids can cause considerable damage to stored seeds. For control of bruchids, see 'Bean'.

Options for improvement
Chickpea seeds can be grown very easily. Varieties do not degenerate and seeds store well. Only in areas where seed-borne diseases are rampant, disease-free seed may find a ready market. In such cases, general extension messages on the identification and control of these diseases can be very effective.

Cowpea

Other names	Black-eye pea, *harocot dolique, kundi* (Swahili), *niebe* (W. Africa)	Market potential seed	poor
		Ease of seed production	easy
Propagation	mainly self-fertilizing		
Storage seed	good		

General description
Cowpea (*Vigna unguiculata*) can be an erect, semi-erect, trailing or climbing legume. Bushes reach heights of 30–90cm. Cowpea is a mainly self-fertilized plant, but insect pollination does occur, especially in more humid conditions, thus increasing the necessary isolation distances from other varieties. Varieties differ greatly in seed colour, but most have the black 'eye' in common.

Agronomic requirements
Cowpea withstands high temperatures and is relatively drought tolerant compared to Phaseolus beans. Cowpeas are often inter-cropped with maize, sorghum or millets. Dwarf varieties can be planted in rows 45–60cm apart, at 10–15cm within the row (population 135 000–150 000 plants/ha). The population of trailing and tall varieties may be less (down to 36 000). The seed rate is 15–30kg/ha. The most important soil fertility factor is phosphorus, which is mostly translocated to the seed. Soils have to be well drained and moist at the time of sowing. The maturity period ranges from three to five months.

Weed control is vital: in most varieties at least two weedings are required, but for trailing types with strong vegetative growth one early weeding may be sufficient.

Harvesting may be done several times for varieties sensitive to shattering. Single harvesting is done when two-thirds of the pods have turned brown. Well-dried pods can be easily threshed without risks to seed quality. Seed moisture should not exceed 13 per cent. Although cowpea is not as severely attacked by diseases as many other legumes, the following diseases need to be dealt with: Cowpea Wilt (*Fusarium oxysporum*), anthracnose (*Colletotrichum lindemuthianum*), scab (*Cladosporium vignae*), and brown blotch (*Colletotrichum spp.*). Insect pests include bean fly (*Melanagromyza*) and pod-sucking bugs (*Nezara*). Intercropping with maize, sorghum or millet may reduce the insect incidence; alternatively, chemical seed treatment (bean fly) or dimethoate field spray (bugs) can be effective.

Seed production

Selection
Selection is quite simple: from a uniform crop, seeds of the right colour are selected. When a mixture has to be purified, line selection is very effective. Major selection criteria are plant type (erect, semi-erect, trailing, climbing), uniform ripening, and seed type.

Seed-transmitted diseases
○ Cowpea mosaic virus: clear mosaic, mottle and green vein banding; causes severe stunting in conjunction with other viruses. It is seedborne,

225

but occurs in some varieties much more than in others. It may also occur in soybean and groundnut (so where this disease is prominent, rotation with these crops is to be avoided).

o Strains of anthracnose (*Colletotrichum lindemuthianum*), angular leaf spot (*Phaeoisariopsis griseola*): see 'Bean'.

Storage
As with Phaseolus beans: a wide range of bruchids are responsible for damage to stored cowpeas. Storage in the pods, mixing threshed seeds with sand, ash, rice husk, and vegetable oil (5–10 ml/kg), or in airtight containers are all effective to some extent. Actellic is a cheap and relatively safe insecticide.

Options for improvement
Cowpea seed production cannot be commercialized easily. The seed stores well, the variety does not degenerate easily, and seed-transmitted diseases are less rampant compared to Phaseolus beans. Improvement of seed quality can best be achieved by general extension messages, particularly aimed at reducing insect damage in storage.

Groundnut

Other names	Peanut, *arachide*, *cacahuete*, *mung-phali*, *mani*, *amendoim*	Market potential seed	fair
Propagation	self-fertilizing	Ease of seed production	easy
Storage seed	poor		

General description
Groundnut (*Arachis hypogaea*) is an erect or trailing plant, with flowers originating in leaf axils. The pegs elongate after pollination and the fruits develop in the soil, 2 to 7cm deep. Pods contain two to six seeds, having a very thin seed coat. Seeds of Virginia type (runner) groundnut varieties exhibit a clear seed dormancy. The crop has a wide variety of uses: consumption of the roasted seeds (runner types), use in sauces and pastes (both runner and bunch), and for extracting oil (Spanish or bunch types), leaving a nutritious cake for livestock feed.

Agronomic requirements
Groundnut is fairly drought tolerant, but frequent rains are necessary during the flowering and fruiting period. Sunshine during the ripening period is necessary for a good harvest and high seed quality. Groundnut is one of the more valuable field crops in most areas. Yields are limited and the amount of labour required for planting, ridging, lifting and plucking is high.

In order to reduce yield losses when lifting the crop, a light and friable soil is necessary. When soils are heavy and rainfall high, planting on ridges is preferred. Good soil preparation is necessary. Whichever planting and harvesting method is used, groundnut is notorious for its volunteer plants which emerge in the season following a groundnut crop. Crop rotation is therefore a necessity for the production of uniform varieties. Groundnut is commonly grown under poor soil fertility.

When planted in rows, inter-row spacing of 40–60cm and within row spacing of 10–15cm are common for bush types (population 140 000 plants/ha.). When planted in the flat, high plant density is possible. Low plant populations generally result in lower Rosette virus incidence, which would otherwise be devastating. The seed rate is 70–110kg/ha. For runner types the spacing is 60 × 30cm, or 55 000 plants and 40kg seed per ha. Groundnut responds well to calcium, potassium and sulphur fertility. It extracts phosphorus from the soil more efficiently than many other crops, thus reducing the effect of applied phosphate fertilizers.

Being a short plant with a rather slow initial growth, weeding is of prime importance for the success of a groundnut crop. Hoeing becomes very difficult after pegging, and weeds have to be pulled by hand.

Most bunch varieties mature in 90–110 days, runner types often after five months only. Plants can be lifted when the seed coat changes colour from white to brown or red. The timing of harvesting is critical. Early harvesting results in shrivelled seeds and poor yield. Late harvesting results in pods remaining in the soil and seeds germinating in the pod (bunch types). The solution is lifting quite early and drying the complete plants. When drying the plants, pods should be off the ground, i.e. ideally placed on racks, otherwise bunches of plants can be placed upside down on the ground. Rain during this weathering period is very damaging. Shelling percentages range from 60 to 80 per cent.

Seed production

Selection
Being a strictly self-fertilizing crop, the selection of groundnut varieties is easy. Removing off-types or a simple line selection are easy to perform and effective. Selection criteria include plant habit, number of seeds per pod, pod shape, and seed coat colour.

Seed-transmitted diseases
Groundnut is attacked by a number of virus diseases, such as peanut stripe (stripes on young leaves, mosaic on older ones), peanut stunt (pronounced stunting), peanut mottle (mottling of young leaves, cupping of older ones) and peanut clump (stunted plants, young leaflets with light green rings).

There is a small chance of transmission through seed but the majority of infections are caused by insect transmission.

Fungus diseases include:

o Early leafspot (*Cercospora arachidicola*), causing circular spots on the leaf, dark brown at the upper side, light brown below.
o Late leafspot (*Phaeoisariopsis personata*), causing small circular spots on the leaf: dark brown at the upper side, nearly black below.
o Rust (*Puccinia arachidis*), causing orange-coloured pustules on the lower side of the leaf.

Finally, groundnut can be attacked by bacterial wilt (Pseudomonas solanacearum) which is a very important disease in potato, tomato, and chili. For all these diseases, seed transmission has been reported. The most important source of the diseases is, however, plant debris and volunteer plants. Proper rotation, the destruction of plants and empty shells, and weeding of volunteers in neighbouring fields are very important. In commercial groundnut seed production, seed treatment fungicides such as Thiram are currently used to reduce seed-borne spores.

Shelling and storage
Groundnut seeds store quite well when dried properly. Seeds generally keep better in the shell. Drying in the shell reduces the drying temperature and thus possible damage, but larger drying floors are required. Without the shell the very thin seed coat may be easily damaged when seeds rub against each other or against the bagging material. This sensitivity also determines the shelling requirements. When shelled mechanically, high moisture-content groundnuts will easily crush; but shelling over-dried groundnuts will cause damage to the seed coat. The nuts may therefore even need to be wetted before shelling. Hand shelling is a very effective means of getting high-quality seed. Stored groundnuts can be attacked by various insects, such as the rust-red flour beetle (*Tribolium castaneum*), the khapra beetle (*Trogoderma granarium*) and especially the West African groundnut seed beetle (*Caryedon serratus*). In the relatively small quantities used in household or village seed production, regular inspections for these insects are feasible.

Options for improvement
Groundnut is an easy crop to produce seed from by farmers who have experience of growing the crop. Varieties do not degenerate because the crop is strictly self-fertilizing; neither seed-borne diseases nor viability after storage are severe problems. Some farmers may need minor technical advice to increase their seed quality.

The reason that groundnut seed may have a ready market is more a matter of availability than seed quality. There is strong pressure for farmers to sell excess groundnuts in the market, since it is a high-value product. Many farmers thus have problems saving enough seed when unexpected expenses arise, such as weddings, hospital bills and school fees. This is particularly pressing, since such a large proportion of the crop has to be saved for seed. With common seed yields of 800kg/ha and a seed requirement of 100kg, more than 12 per cent of the yield has to be stored, which is higher than in any other seed crop. General shortages of groundnut seed are therefore common. Assistance should pay attention to this aspect.

Finger millet

Other names	African millet, *ragi* millet, *kurakkan*, *dagussa*	Market potential seed Ease of seed production	poor easy
Propagation	self-fertilized		
Storage seed	good		

General description
Finger millet (*Eleusine coracana*) is a short, profusely tillering plant with characteristic finger-like terminal inflorescences, bearing small reddish seeds. The crop matures in three to six months depending on the variety and on the growing conditions (especially temperature). The crop is adapted to fairly reliable rainfall conditions and has an extensive but shallow root system. The crop is important in semi-humid climates in Africa as long as the soils are well drained.

Agronomic requirements
Because the seeds are small, seed-bed preparation has to be done with care. A fine tilth is necessary to produce an even germination, which is important for the early suppression of weeds. Finger millet is commonly broadcast at 10 to 15kg seed per ha. When drilled in rows, row spacing of 25cm is common. Being relatively short, finger millet may suffer badly from weeds.

The crop grows best on sandy loam soils. Even though often grown on rather marginal land, finger millet responds well to fertilizer applications or residual fertility from well-fertilized and weeded crops like tobacco, groundnut or soybean.

The crop does not suffer from many pests. Only occasional outbreaks of army worm and bird damage occur.

Seed production

Selection
Finger millet is a self-fertilized species of which quite uniform varieties exist. Reasons to select for uniformity are ease of weeding (uniform plant

height) and harvesting (uniform maturity). Isolation requirements are minimal and purification of varieties straightforward (negative mass selection or the ear-to-row method).

Weeds and diseases
The plant population is a major concern in finger millet seed production. Weeds get a chance at low plant density and can reduce yield and subsequent seed vigour problems, and grassy weeds can also produce seeds of similar sizes that are difficult to remove. The most noxious weed is the closely related species *Eleusine indica*. Row planting helps weeding considerably and higher seed rates at planting may reduce the weed problem to some extent.

Seed-transmitted diseases include blast (*Gloeosporum*) and leaf blight, which can be controlled with chemical seed treatment, but major disease problems are not common.

Harvesting, cleaning and storage
Tillering compensates for low plant populations, but extensive tillering increases the problem of uneven ripening of the heads, resulting in a large proportion of immature seeds at harvesting. Several rounds of hand picking of ripe heads produces the highest yields of quality seed, but is very labour intensive.

Cleaning, and especially the winnowing operation, is very important to remove the immature seeds. Millet seed stores well compared to most other cereals. A seed moisture content of 12 per cent is sufficient.

Options for improvement
Finger millet is a relatively easy-to-produce seed crop, apart from standard cultivation problems like weeding and harvesting by hand. Farmers who are used to the crop generally do not have problems in producing their own seed. Since finger millet is grown in quite benign ecological areas and since small quantities of seed are required, seed security is hardly ever an issue. Seed enterprises based on finger millet seed are unlikely to emerge. Where farmers do face problems with finger millet they may relate to the variety. The purification of existing varieties or the introduction of new ones may be regarded as valuable forms of assistance.

Maize

Other names	corn, *mais, maíz, milho, makkai, cintli*	Market potential seed	good
Propagation	cross-fertilized (wind)	Ease of seed production	fair
Storage seed	fair		

General description

Maize (*Zea mais* L.) is an erect plant of 1.5–5m height that forms very few tillers. Some tiller production may be favoured in case of multi-purpose maize for both human food and animal feed/forage. Maize is normally cross-fertilized by wind-carried pollen. The ear, which is formed about mid-way up the stem, contains 8–28 rows of grain, each row having between 20 and 60 grains. Grain colour can be white, yellow, red or blue, and grain type varies between dent (soft) and flint (hard). Special-purpose maizes (e.g., pop corn, baby corn, sweet corn, quality protein maize) serve niche markets. The maturity period ranges between three and five months depending on the variety and environmental conditions (day length and temperature). Most maize varieties are photoperiod sensitive. Care must therefore be taken with introduced germplasm. Most commercial and an increasing number of public varieties are hybrids. Germplasm is loosely grouped in lowland tropical, sub-tropical, temperate and tropical highland categories. Apart from their adaptation to temperature and length of season, they also carry particular disease resistances. When grown as a mono-crop, rotation is advised with legumes, or other unrelated crops, such as potato, cotton or sugar cane, but rotation schemes with other cereals (sorghum, wheat, rice, barley) can also be successful.

Agronomic requirements

Maize is adapted to a wide range of conditions and is cultivated from the lowland tropics to temperate conditions. Maize exhausts the soil and is very responsive to well-prepared soil of high organic matter and soil fertility. Good growing conditions are particularly important during flowering (when the silk is fresh). Limited moisture, low nitrogen, water logging, low light intensity, and too high plant density may cause kernel abortion and reduction in ear number. This is indicated by a delay in silk extrusion. Temperatures above 35°C cause pollen sterility. Tall varieties have to be sown at reduced plant density. The plant height makes the crop susceptible to lodging, especially when winds are strong and the soil very wet. A minimum rainfall of 500mm during the growing season is needed to reach an acceptable production. Nitrogen fertilizer is best given as top dressing mid-way and at flowering stage (but in limited amounts when grown for seed). Intercropping with a legume is common practice. A row distance of 70 to 100cm facilitates weed control, and enables access to this tall crop for visual inspection of individual plants and hand-pollination. Plant populations of 36 000 plants per hectare (spacing of 90 × 30cm) are common for tall varieties, but may go up to 70 000 (70 × 20cm) for shorter types in high rainfall areas. Seed use is 20–35kg/ha depending on seed size and desired population density.

Borers and fall army worm (*Spodoptera ssp.*) are difficult to control once they have entered the crop. The removal of stubble, crop rotation and preventive spraying before flowering are the major means to reduce damage.

Important diseases include maize streak virus, southern rust (*Puccinia polysora*), leaf blight (in cool areas, northern leaf blight, *Helminthosporum turcicum*, and in warm areas southern leaf blight, *Drechslera maydis*), downy mildew and ear rots. The incidence of most other diseases may be reduced by crop rotation and timely destruction of the stubble.

Maize can be associated with parasitic weeds (e.g., *Striga* in Africa). If ears are heaped on the bare ground during harvest, weed seeds may be attached to the seed.

Seed production

Selection
Since maize is a cross-fertilized crop, plants producing seed should be isolated from other maize varieties. The outer rows facing the other varieties should be harvested for consumption whereas the centre can be used for seed. Selection is carried out: before flowering for resistance to slowly developing foliar diseases and good tolerance to insects; around flowering for average anthesis date and normal tassel (male flower); after flowering for sturdy stems that do not lodge, low or no tillering, high tolerance to foliar diseases, low ear and plant height, good husk cover and general plant type; and after harvesting for ear characteristics (size, flint/dent, colour, ear rots, insect damage). There is a danger with selection for large ears after harvest: it may imply indirect selection for tall, later-maturing plants. Selection therefore preferably involves marking of good plants in the field during the season. A minimum of 300 ears, but preferably 500 ears, should be selected to avoid genetic drift.

Seed production from a commercial hybrid variety of maize is not recommended. Farmers may obtain a reasonable yield with such populations, but generally only after a number of cycles of selections aimed at improvement of performance, uniformity and adaptation. However, that effort may be worthwhile if the hybrid contains valuable characteristics. Local seed production from composite or synthetic varieties is possible.

A large germplasm collection is maintained by the International Maize and Wheat Improvement Center (Centro Internacional de Mejoración de Maíz y Trigo) near Mexico City, Mexico. Many national germplasm banks also possess collections.

Seed-transmitted diseases
Major seed-transmitted diseases are:

○ Ear and stalk rot (*Diplodia* and *Fusarium* fungi) and storage rot (*Aspergillus*) are especially common after wet weather late in the season.

Control measures are crop rotation, good potassium fertility and avoidance of high nitrogen levels, and seed treatment with common fungicides (captam, thiram, etc.).

○ Southern corn leaf blight (*Drechslera maydis*), characterized by long streaks of spots scattered over the leaves, which turn brown and die. It is common in warm climates and should not be confused with northern corn leaf blight (*Helminthosporium turcicum*) which is not seed transmitted and occurs in cooler areas (above 1000m in the tropics). Crop rotation and seed treatment (captan, thiram and carboxin+thiram) are major control measures.

○ Downy mildew (various *Peronospora/Sclerospora* species). Most diseases of this group are characterized by leaf streaking and stunting, and are confined to Asia and the Americas. It should not be confused with maize streak in Africa, caused by a non-seed-transmitted virus or with soil deficiencies. Control by early roguing or metalaxil treatment.

○ Common smut causes large galls protruding from the husk. Spores can be carried with seed. Control is possible through early removal and burning of infected plants (roguing) and carboxin+thiram seed treatment. (In Mexico, however, these 'Juitlacoche' fungi fetch a high market price as a speciality food).

Storage

Maize is harvested when the seeds are hard and glazed. Freshly harvested seeds contain up to 30 per cent moisture. Drying can be done in specially constructed maize cribs or on racks, and shelling can be done at 22–25 per cent moisture. Further drying can be done on concrete floors in the sun. When storage space is not limiting, it is advisable not to shell and store the seeds on the cob. Maize seed at 12 per cent moisture content can be stored for prolonged periods in airtight containers (see Section 7.3), or well-ventilated containers or structures that have a relative humidity of less than 60 per cent and temperatures of up to 30°C.

Aspergillus flavus is a dangerous fungus that may develop on moist maize. It produces aflatoxin, which is hazardous to humans and animals. Its consumption should at all times be avoided. Whether stored on the cob or shelled, maize is threatened by various grain weevils (*Sitophilus oryzae, S. zeamais*). Infection occurs earlier in the field. The ear borer (*Mussidia nigrivenella*) attacks the seed between maturity and drying. The larger grain borer (*Prostephanus truncatus*) attacks in the storage place. *P. truncatus* is found in Africa and the Americas, and attacks other crops as well. These insects can be controlled by (1) rapid drying, (2) hanging the cobs over a fireplace, (3) the use of insecticides or local plant extracts, or (4) shaking the container with shelled maize regularly to extend the reproduction cycle of the insect. The main factors that favour insect infestation are oxygen availability (very well-dried seeds can be stored in air-tight containers), high air humidity and temperature.

Options for improvement
Maize seed can relatively easily be produced by individual farmers. Farmers generally select good ears and store these in their kitchen or under a shed. There can also be a fair demand for well-selected maize seed with good ear characteristics and high genetic uniformity with respect to seed colour and type, and possibly treated with fungicide, as preferred by the commercial market. In areas with severe fall army worm or stalk borer presence, insecticide treatment may be demanded. Maize offers a fair potential for improvement of both local and modern OP varieties and can be a major product in local seed enterprise development.

Rice

Other names	*riz, arroz*	Market potential seed	poor
Propagation	self-fertilizing	Ease of seed	
Storage seed	good	production	easy

General description
Rice (*Oryza sativa* L.) has three ecogeographical sub-species: *indica* is cultivated in the humid tropics, has thin, long grains and is mostly adapted to short days; *japonica* is cultivated in sub-tropical and temperate zones, has round to oval grains and possesses a low or no day-length sensitivity; and *javanica*, which is cultivated in parts of Indonesia. Other rice species are found in West Africa: *O. glaberrima* and *O. breviligulata* are perennial, but are mostly grown for only one harvest.

The tropical types require a growing season of three to six months with optimal 32°C and not below 20°C. Because of this, tropical rices are seldomly grown at higher altitudes.

Irrigated rice is usually sown in nurseries, from which seedlings are transplanted to inundated fields. Upland rice is usually direct-sown. Rice is a strongly tillering crop of 0.5–1.5m height, with the major tillers producing productive panicles of 15–40cm length and with about 100 kernels (wide variation exists). The development of early maturing varieties, in combination with an uninterrupted water availability during the year, has enabled the cultivation of 2–3 crops per year. Pests and diseases can cause serious grain yield reduction. The crop's self-fertilizing nature facilitates seed multiplication. Hybrid varieties are extensively used in China.

Agronomic requirements
The water requirements of rice (including upland rice) are high, which limits its spread to regions where water is not abundantly available from

rainfall or irrigation. The minimum seasonal requirement is about 800mm, and 1250–1500mm is considered the optimum. Although rice produces fair yields under low management conditions, especially short varieties respond well to applied nitrogen. Split application is effective to reduce leaching. Phosphorus, to be applied at or before planting, is necessary for good root development. All soil types are acceptable. The oxygen level in the soil should not reduce too much, as this will limit nutrient uptake. This may be caused by high water levels and high temperatures. The crop lodges easily. Rice is a short-day plant, but modern, photoperiod insensitive varieties have been developed.

Seedlings are transplanted to the rice fields about three weeks after emergence. Plants are transplanted at a rate of 2–5 plants per hill, depending on tillering capacity and agronomic practices. Hill distance is 10 × 10cm to 20 × 20cm. It is important to avoid the formation of too many non-panicle producing tillers, however, a moderate level of tillering may be desirable for the compensation by additional tillers that may be needed in case of stem borer infestation of the main tillers.

Pests and diseases that can cause severe yield reduction include blast (*Pyricularia oryzae* Cav.), bacterial leaf blight (*Xanthomonas oryzae*), stem borers (*Chilo* spp., *Sesamia* spp., *Tryporyza* spp.) and leaf hoppers (*Sogatodes* spp.). Control of these can be very problematic. Many other pests and diseases occur in different countries. Also rodents, hoppers and snails can cause severe damage. There are three approaches to weed control in irrigated rice: cultural, mechanical and chemical. Cultural control includes planting weed-free seeds, proper seed-bed preparation, land levelling, levee construction, water management and rotation (most effective). As the seeds begin to mature, water is drained from the field, so the seed can dry and the field can be entered easily for harvest. Various weeds can cause severe problems in rice production, *Rottboellia* being one of the most noxious species. Harvesting is usually done by hand, using a small knife or a sickle, when the moisture content of the kernels is about 20 per cent. Further drying is done by stacking panicles on a rack, which is laid on the ground. Threshing is done with more or less sophisticated mechanical equipment, or with simple devices such as a lout.

Rice is mostly grown in monoculture, and can be farmed in crop rotation with sugar cane, cotton, legumes, or wheat.

Seed production

Selection
Rice is a self-fertilizing crop, that shows strong fertility barriers between subspecies, but not between cultivars. A distance of one or a few metres, between different cultivars is recommended. Rice varieties are commonly

quite uniform and being principally self-fertilizing allows for relatively easy selection when necessary. Taller off-types are normally easily recognized, but shorter ones often remain unnoticed.

A large germplasm collection is maintained by the International Rice Research Institute in Los Baños, the Philippines. Many national germplasm banks also possess extensive collections. Such genebanks can be important in providing materials for participatory variety selection or plant breeding.

Red rice, which shatters easily, and which can be difficult to distinguish from the cultivated rice, can best be removed manually from the field before it shatters, as it will otherwise contaminate seed lots. The best way to identify such off-types is to plant the crop in rows. In addition, weed problems, including red rice, may be reduced with special methods of land preparation, such as accurate levelling and double preparation (allowing weeds to germinate and killing them off in the second round).

Seed-transmitted diseases

Some of the major rice diseases are seed transmitted and should thus be handled with more care compared to standard paddy production. The following fungal diseases are seed transmitted: bunt (*Neovossia horrida*), smut (*Ustilagionoidae virens*), grain spots (*Nigrospora sp.*) and scab (*Gibberella sp.*). Treatment with Thiram or Captan, possibly in combination with Vitavax, may be considered. The most widely feared rice disease, blast (*Piricularia oryzae*), is, however, not seed transmitted. In addition, bacterial blight (*Xanthomonas sp.*) can be seed transmitted and should be guarded against.

Storage

Rice seed is harvested at a moisture content of about 20 per cent, and must be dried quickly to avoid heating, discolouration, and a loss of viability and vigour. Sun-drying for 4–5 days is sufficient. The drying temperature should not exceed 35 °C, as this causes seed cracking. Rice can be stored for a number of years if its moisture content is below 10–12 per cent.

Options for improvement

Rice seed can easily be produced by competent rice farmers. The commercialization of rice seed production is therefore rather difficult. Improvement of the rice seed thus normally involves the selection of better varieties or the purification of existing ones. In some cases it is worthwhile to reintroduce traditional varieties which have particular quality characteristics and thus fetch a better price in the market, which can compensate for a lower yield.

Improvement of varieties usually focuses on yield level and stability, and related disease and pest tolerances. Purification of populations of mixed

236

genetic background by pure-line selection is usually the first and easiest step to improve characteristics. As rice does not show inbreeding effects, the elimination of a majority of lines and maintenance of only a few good ones does not lead to yield depression as a consequence of genetic narrowing. However, it is advisable to maintain a certain degree of genetic heterogeneity if there is reason to believe that this improves tolerances to pests and diseases, and yield stability. Further improvement may be attempted by hybridization between individuals from the same sub-species.

Pearl millet

Common names	*dukhn* (Arabic), *bajra* (Hindi), *millet chandelles, millet perle* (French), *mijo peria* (Spanish), *milheto messango* (Portuguese), Indian millet, bulrush millet, *mwele* (Swahili)	Market potential seed	poor/ hybrids good
		Ease of seed production	quite easy
Propagation	cross-fertilized		
Storage seed	fair		

General description
Pearl millet (*Pennisetum americanum* (L.) Leeke) is an annual erect plant of up to 4m high with solid, round or oval stems and slightly swollen nodes. The leaves are arranged alternately along the stem and are glabrous or more or less hairy. The inflorescence is a terminal spike-like (column-like) panicle carrying large numbers of small round seeds. The crop is cross-fertilized. The crop commonly matures 3–4 months after planting.

Agronomic requirements
Pearl millet grows well in the drier parts of the hot tropics, especially in southern and Sahelian Africa and the Indian sub-continent. Very hot weather during flowering, however, may reduce the seed set. Sowing is best done in rows 50 to 150cm apart (depending largely on the available moisture) in a fine seed bed. Weed competition can be severe in a young crop, but once established, weeds only compete with pearl millet for moisture. Pearl millet can be attacked by Striga (parasitic weed).

Seed production

Selection
Pearl millet is a cross-fertilized crop and very heterogeneous in most farming conditions. Hybrids have become very popular in large parts of India but hardly at all in Africa. Improvement may be possible through selection

within existing populations. Early vigour, good tillering, strong stems and heads with bristles (against birds) may be important selection criteria. Strong negative mass selection before flowering may reduce the percentage of clearly off-type plants. Otherwise, half-sib selection with the remnant seed method may prove useful (see Section 8.2.5). Being a cross-fertilized crop, care should be given to avoid inbreeding and genetic drift by selecting sufficient numbers of plants.

Seed transmitted diseases
A variety of diseases attack pearl millet. The most common are:

○ Green ear (*Sclerospora graminicola*), causing yellow streaks on the leaves and downy growth at the lower surface in damp weather. The head is transformed into green leaf-like structures instead of seeds.
○ Ergot (*Claviceps fusiformis*) causes honeydew to develop in the heads. The seeds are covered in purple-black sclerotia. This disease is both soil and seed borne.
○ Smut (*Tolyposporium penicillariae*) also transforms florets in fungus-balls which turn from green to dark brown. Seed treatment with Captan or Thiram, possibly with added Vitavax, is effective.

Harvesting and storage
Pearl millet should be harvested at a seed moisture level of about 22 per cent. Manual harvesting is common and is particularly recommended with open-pollinated varieties that generally have a variable ripening. The heads can be dried on racks or drying floors until the seed reaches 12 per cent moisture content. Rains during harvesting or drying may cause early germination of the seed on the head. Seed can be stored on the head or threshed.

Options for improvement
Pearl millet varieties can be very heterogeneous. (Participatory) variety selection and breeding, including purification of existing landraces may be a very effective means to increase the yield of the crop and quality of the seed. Since pearl millet is grown in very marginal conditions, seed security may be a serious issue. Introducing pearl millet seeds from distant sources during emergency seed campaigns may, however, be very disappointing.

Chemical seed treatment can reduce the damage by important diseases. Since in most areas in Africa pearl millet is basically a subsistence crop, commercial seed supply is very limited. Developments in countries such as India, however, show that pearl millet can be a very good crop for small-scale seed business development.

Potato

Other names	Irish/English potato, *pomme de terre*, *papa*, *batata*	Market potential seed	good
		Ease of seed production	difficult
Propagation	vegetative (tubers); sometimes seed		
Storage seed	not easy		

General description

Potato (*Solanum tuberosum*) is an erect herbaceous plant, normally grow-ing from a planted tuber. A planted tuber develops one or several stems which grow from the 'eyes' on the tuber which may branch above or under ground. The primary stems behave as individual plants with their own root system. Plant population is therefore best expressed as the number of stems per hectare instead of the number of planted tubers. Crop maturity strongly varies, depending on the variety and the growing conditions (tem-perature, day-length, soil fertility). Average yields are 15 t/ha in Peru and 50 t/ha in the Netherlands, with a global average of 14 t/ha.

A stem can develop one or more inflorescences, and this is stimulated by long day conditions. In some varieties, flowers do not develop at all. Flowering may be stimulated by removing the developing stolons and tubers.

Underground parts of the stems can develop into stolons, i.e. horizon-tally growing branches that develop tubers at their tips. Tuberization may respond to day-length and temperature, and commonly starts 4–6 weeks after emergence. Tubers are usually dormant directly after harvesting, i.e. they do not develop sprouts even when exposed to favourable conditions. The dormancy can last 1–4 months depending on the variety, the growing and the storage conditions. Commonly, only the top eye will develop a sprout. More sprouts develop when the dominant sprout is getting old, when it is removed or after a prolonged storage of the tuber. In the latter case the many sprouts that develop may, however, be weak. Planting tubers with three to six sturdy sprouts is the optimum for a good crop.

Cool storage delays the ageing of the seed tubers. Some local storage constructions, such as '*narwallas*' in Egypt and others allowing diffuse light keep seed tubers in good condition for a relatively long period. Diffuse light slows down the degeneration process and produces short and sturdy so-called 'lightsprouts'. In other conditions artificial cooling may be needed, which is expensive. Seed tubers stored in the dark develop long slender sprouts that easily break during transportation and planting. These should therefore be removed 2–3 weeks before planting to allow new sprouts to emerge. If unsprouted tubers are planted, emergence may be delayed and rotting may occur.

Agronomic requirements

Potato is a crop of moderate temperatures and in the tropics is mainly found in mountainous areas, even though varieties are being bred that do well under warmer conditions. A night temperature below 18 °C is needed for tuberization; above 22 °C there is hardly any. Day temperatures of 20–25 °C are optimal. Potato requires moderate, well-distributed rainfall (500–750mm) and a quite high soil fertility tolerates a wide pH range (4.8–7). It does not tolerate heavy waterlogged clay soils.

Tubers of 20–50mm diameter are usually planted. Smaller ones produce fewer and slower-developing stems. The use of larger seed tubers has no agronomic advantage and means higher cost (storage, transport) and opportunity cost (food value). Tubers are normally planted in rows at a depth of 5–15cm, measured from the top of the tuber. In monocropping for ware potato production rows are 70–100cm apart at a spacing of 30–40cm in the row (25 000–40 000 tubers/ha). In specialized seed tuber production up to 80 000 tubers/ha are planted with the same row-width. Depending on the size of the seed tubers and the planting density, 1.5–4 ton/ha are used for planting.

Earthing-up or hilling is done 3–4 weeks after emergence to control weeds and to avoid greening of the growing tubers.

Seed production

Selection

Since potato is vegetatively propagated, all tubers from one variety are genetically identical. Local varieties may be a mixture of similarly-looking genotypes, and in some countries widely differing potato clones are mixed in one field.

Potato is a tetraploid, and using seed to improve varieties will not readily produce acceptable new varieties because of wide segregation. Various *Solanum* species are grown for their edible tubers in the Andes and are important sources of disease resistance in formal breeding programmes. The International Potato Centre (CIP) is a major source of genetic materials and information on the crop.

Seed-transmitted diseases

Production of quality seed tubers is difficult, especially in hot humid environments, because of the many diseases that are transmitted through vegetative propagation. Moreover, insects and diseases spread readily during storage and in the field. Rotation is important to prevent a build-up of pathogens. The most important diseases are:

o bacterial diseases: bacterial wilt (*Pseudomonas solanacearum*), and soft rot (*Erwinia carotivora*). Common scab (*Streptomyces*) affects the skin, but does not seriously reduce yields.

240

o fungal diseases: late blight (*Phytophthora infestans*), early blight (*Alternaria solani*), black scurf (*Rhizoctonia solani*) and pink rot (*Phytophthora erythroseptica*).
o viral diseases: potato leaf roll virus (PLRV), potato virus X, potato virus Y and mosaic virus A are the major yield-reducing viruses. Most viruses are transmitted by aphids, through leaf contact and tubers. Tuber infection is commonly invisible, but a trained eye can spot virus on leaf symptoms such as rolled lower leaves, mosaic patterns, and reduced plant size.
o nematodes: especially cyst nematodes (*Globodera spp.*). Resistant varieties do exist, but rotation appears in most cases to be the main option, since chemical control is very expensive.
o aphids, tuber moth and spider mites. Tuber moth can be a serious problem during storage and may be controlled by the use of a granular product carrying a virus that kills the larvae, and by careful hilling and harvesting. Pest damage may be reduced through early planting.

Harvesting, processing and storage
Tubers that are to be used for seed are commonly harvested early to avoid tuber infection with aphid-transmitted viruses and late blight. Particularly the early harvested tubers have to dry for 1–2 weeks to cure the skin. A seed lot should not contain diseased tubers since diseases may easily spread in storage. Ideally, seed tubers are stored in crates in a cool and rat-proof place where diffuse light can enter. Low-cost stores have been designed by CIP.

Alternative propagation methods
Stem cuttings, leaf cuttings and sprouts can be used for planting, but require special care. Botanical seeds can also be used. This 'true potato seed' (TPS) technology has been widely explored during the past decade particularly, because the many tuber-transmitted diseases are not seed borne. The true seed can be extracted from the berry in the same way as for tomato, after hand pollination of selected parent plants. Seeds from spontaneously developing (selfed) berries generally show much segregation for plant size and tuber characteristics. Potato seeds are normally dormant for 6–12 month after harvest. Growing the seedlings requires careful nursery management: soils need to be high in organic matter, and frequent watering, fertilization and hilling are needed (50–75 seedlings per m^2). The seedlings develop small 'tuberlets' that can be used as seed tubers or to grow a ware crop. The small nursery area allows for off-season production of the tuberlets.

Options for improvement
In tropical lowlands it is very difficult to keep the seed tubers free from diseases. It is therefore more effective to support farmers in higher altitude

areas to improve their seed quality and subsequently to develop business contacts with the lowlands. Identification of the many diseases is an important step in the production and selection of healthy seed. Also, the improvement of the (diffuse light) storage conditions and the testing of disease resistant varieties can be an effective means of assisting seed producers.

Potato is a very good seed crop for commercialization when quality problems can be solved. Because of the distances involved and the bulkiness of the product, relatively large investments are needed.

Sesame

Other names	simsim, beniseed, gingelli, sesamo, til	Market potential seed	fair
Propagation	semi-cross-fertilized by insects	Ease of seed production	easy
Storage seed	good		

General description

Sesame (*Sesamum indicum*) is an erect branching hairy plant, 60 to 120cm tall. Fruits develop in leaf axils and contain many small seeds. The crop is adapted to low rainfall areas (500mm during growing period), but it does not stand severe drought or water logging. Flowering of the plant depends on the day-length; varieties may differ in this aspect. The white or pink flowers are visited by insects, but because the stigmas grow through pollen-shedding anthers, most of the flowers are self-fertilized. Plants flower from the bottom upwards. As a result there are mature fruits at the bottom of the plant and immature ones at the top at the end of the growing period.

Agronomic requirements

The seeds being small, a well-tilled seed bed is essential. Timing of planting is essential, because rain should never interfere with harvesting. Row planting is recommended for seed production. Row widths of 30–60cm are used in dribbling 3–5kg seed per ha. Thinning is often necessary to arrive at 10–15 cm between plants in the row (population approximately 200 000 per ha). The seedling stage is very delicate and moisture stress, water-logging and weed competition can seriously affect the young crop. Most local varieties of sesame are dehiscent, i.e. the seeds shatter very easily. Early harvesting when the lower fruits turn yellow is necessary to avoid serious yield losses. Plants are then commonly hung upside down on slanting racks to dry, while the upper fruits ripen further on residual plant moisture. Many modern varieties are non-dehiscent.

Seed production

Selection

Since sesame attracts insects, a considerable amount of cross-fertilization may occur in this basically self-fertilizing crop. Sesame may therefore be called semi-cross-fertilized. This means that a high degree of varietal uniformity can be obtained, but that genetic degeneration will quickly occur when the variety is not well maintained.

Especially for the high-value confectionery sesame, where the market prescribes pure white seeds, considerable effort may have to be put into keeping the variety pure. This can be done through regular line selection, possibly with the remnant seed method. Selection normally concentrates on seed colour, maturity period, reduced branching and even plant height.

The day-length sensitivity is significant when importing varieties from other areas. When the crop is to be grown during spring in the sub-tropics, i.e. when the day length increases with time, varieties adapted to the short day tropical climates may not perform well. Variety trials under local conditions and in the different seasons should be performed before a recommendation can be given.

Seed-borne diseases

Important diseases are different leaf spot diseases, caused by the fungi *Cercospora sesame* and *Alternaria*, and by *Pseudomonas sesame* bacteria. The fungi may be controlled by Thiram seed treatment, but field infection may occur from stubble. Spots occur normally only during flowering. Furthermore, Fusarium wilt may cause withering of young sesame plants.

Storage

Stored sesame is not attacked by major pests, and the seeds maintain their viability quite well, especially since harvesting is normally done under very dry conditions.

Options for improvement

Sesame seed can be grown by farmers quite easily. Commercial prospects for seed production are therefore limited. Since sesame is grown in quite arid areas, seed security may be a problem after a prolonged drought. Storage is, however, without much risk.

The main reason for buying sesame seed is to improve on the marketability of the produce, i.e. to get a uniform seed colour. In areas where brown and black-seeded plants are common, well-selected seed should have a good market for this high-value crop.

Sorghum

Common names	durah (Arabic), jowar (Hindi), sorgho (French), sorgo (Spanish, Portuguese)	Market potential seed	limited for OP's, good for hybrids
Propagation	semi cross-fertilized	Ease of seed production	easy
Storage seed	fair		

General description

Sorghum (*Sorghum bicolor* (L.) Moench) is an annual erect plant of up to 4m high with a dry or juicy stem, and an often hairy sheath covering part of the internode and alternate flat leaf blades. The plant has a stature like maize, but with a tendency to tiller. The terminal inflorescence is a compact, semi-compact or loose panicle. The fruit is a usually naked caryopsis. The plant is usually photosensitive, but insensitive types are commonly cultivated. The traditional varieties are tall and late maturing; improved varieties tend to be shorter and earlier. The taxonomy of sorghum and related species is complex and confusing due to inter-fertility. Five main types can be identified: *Bicolor, Guinea, Caudatum, Kafir*, and *Durra*.

The crop is largely self-fertilized (but also out-pollinated in 5–10 per cent). In some areas hybrids are grown. In the following notes, however, only open pollinated (non-hybrid) seed production is considered.

Agronomic requirements

Sorghum is native to Africa and is grown mainly in areas which are too dry for growing maize. It extends to areas with only 300 to 400mm average annual rainfall. Temperatures of up to 40°C can be tolerated without problems. It can withstand drought, and when rains arrive it again starts growing vigorously, hence the name 'camel crop'. Germination of seeds may pose problems, especially if seeds have not been kept dry enough and mould has developed. Hot and moist sowing conditions are unfavourable. Optimum germination temperature is between 20 and 30 °C with an optimum around 25 °C. Sorghum can grow on almost any soil, but loamy soils are considered to be the best, and the pH may range from 5.6 to 8.5. Planting distances vary from 45 to 90cm between rows, and within the row from 15 to 60cm.

Shootflies cause dead hearts in young plants, resulting in early tillering; stem borers also resulting in dead hearts in young plants, in older plants 'shot holes' appear on the expanding leaves. Other harmful insects are earhead midges (white panicles, pale orange fluid from squeezed florets, no seed setting), earhead bugs (nymphs and adults on panicles, no seeds), caterpillars, grasshoppers, aphids and mites.

Sorghum is attacked by a large number of fungal diseases, including smuts (*Sphacelotheca* spp.), rust (*Puccinia purpurea*), downy mildew

(*Sclerospora sorghi*) and ergot (*Sphacelia sorghi*). Loose smut (*Sphacelotheca cruenta*) causes early heading, dwarfing of plants, and heads transforming into black spore masses; covered smut (*Sphacelotheca sorghi*) causes the seeds to be converted into fungal bodies (sori), the spores remain covered and the plants show no dwarfing or other apparent symptoms. Rust (*Puccinia purpurea*), causes red to purplish mealy spots of various shape on the leaves.

Seed production

Selection
Selection is preferably done in the field: in this way all plant characteristics can be taken into account. Important selection criteria include plant stature (including seed-plant ratio, plant height, number of leaves, number of tillers), relative disease-freedom compared to the other plants, maturation time, seed size, seed colour. At least 100 panicles should be labelled and harvested at maturity in order to maintain the necessary variability of characteristics. If not so much seed is needed for sowing next year, all the panicles should be threshed, the seed mixed and the portion needed kept. This seed will be used as stock seed for next years' seed production; the other seed may be distributed for crop production.

If a more homogeneous crop is desired, a line selection should be carried out. In sorghum some variation will occur due to cross-pollination and heterozygosity. Still, in one or preferably more seasons one can obtain a number of selections in this way. Due to its tendency to cross for 5–10 per cent it must be regarded as a cross-pollinator and an isolation distance of 200m is required. If very different types are grown, a distance of up to 400m is recommended. If varieties are grown which are uniform and short in flowering, time isolation can be considered by staggering the planting of the different varieties, but because of a large variation in flowering within the crop, this practice is not very reliable. Seed fields and surroundings should be absolutely free from wild relatives (*Sorghum halepense, Sorghum verticilliflorum*) because they readily cross.

Seed transmitted diseases
Of the major sorghum diseases the following are seed transmitted: smuts (*Sphacelotheca* spp.), downy mildew (*Sclerospora sorghi*) and ergot (*Sphacelia sorghi*) and sugar cane mosaic virus (SCMV).

The most effective measure to control smuts is through seed treatment with Captan or Thiram. Manually removing diseased plants from the field is only partly effective and should be done as early and thoroughly as possible. But because most plants look healthy until the spores break out, the other grains may already be bearing spores by that time. During threshing and cleaning, the entire seed lot gets contaminated with spores from just a

245

few diseased seeds. Covered smut is seed borne, loose smut is both seed and soil borne, while head smut (*Sphacelotheca reiliana*) is entirely soil borne. Ridomil can be used for treating seed against downy mildew.

Harvesting and drying

Seed should be harvested after it has reached 15 per cent moisture content. This is usually the case when most of the lower leaves have dried and the upper ones have turned yellow. If the climate is dry one can leave the seeds drying on the plant further, in wet climates one has to harvest as soon as possible to avoid mould and sprouting. If there is variation in maturation, one should harvest two or three times. Panicles are cut off manually and left to dry in the field or on drying floors. The layer of seeds should preferably not be thicker than 10cm and should be turned regularly, to obtain an evenly dried crop and to avoid water accumulation in the lower layers.

Processing and storage

To avoid damage during threshing, the moisture content of the seed should be 12 per cent or lower. Seeds of other sorghum species are very difficult to remove; Johnson grass can sometimes be removed with carefully chosen slotted screens or indented cylinders. Other kinds of seed usually deviate so much in shape, size or density that removal with normal cleaning machines is relatively easy. Under farmer's conditions most of the foreign seeds can be blown out during winnowing, because the majority are lighter or more chaffy.

Seed treatments may include chemical dressing against smut and bunt. Storage of sorghum seed is usually not problematic if seed moisture content is equal to or lower than 10 per cent. The seeds can be stored hanging under the house-roof or in a storage shed. Larger quantities are usually stored in granaries (on the panicle) or threshed and bagged. The seeds are vulnerable to insect and rat attack and moulds. Application of an insecticide (e.g. Actellic) is recommended.

Options for improvement

Sorghum seed is commonly produced by farmers themselves. A wide range of landraces has evolved over time, which may be well adapted to the particular ecological and socio-economic conditions of the different production regions. Changing climates and changing markets may, however, warrant improvement of the varieties through (participatory) plant breeding and variety selection. A large germplasm collection is maintained by the genebank of the International Centre for Research in the Semi Arid Tropics (India, Niger, Zimbabwe).

Major local seed production problems include insect damage during storage and seed-transmitted diseases (smut). Both can be solved with agro-chemicals, which may give the sorghum seed a fair market. Where hybrid sorghums can be grown economically, the crop has a ready market for seed entrepreneurs. Furthermore, sorghum is a crop with particular seed security aspects where it is a basic food crop grown in drought-prone areas. The availability of widely differing varieties bears the risk of supplying the 'wrong' sorghum seed in emergency seed provision programmes.

Soybean

Other names	Soya bean, soya, *ta-tou* (Chinese)	Market potential seed	fair/ good
Propagation	self-fertilizing	Ease of seed	difficult
Storage seed	poor	production	

General description
Soybean (*Glycine max*) is an erect annual legume, usually 45–90cm tall with dark green hairy leaves. The very small flowers are almost strictly self-fertilizing. Soybean is a short-day plant. The result is that varieties introduced from other latitudes may not flower at the desired time; and thus have a very different maturity period than expected. Maturity periods range from 2.5 to 7 months for different varieties.

Soybean is one of the world's most important sources of protein and vegetable oil. Unripe seeds are eaten whole or sprouted as a vegetable, but most soybean for human consumption is processed into curds, sauces or many other products, or crushed for oil. The main use world-wide is, however, animal feed (chicken, pigs, cattle).

Agronomic requirements
Soybean is adapted to a wide range of climates, provided water is available in well-drained soils. Soybean exhausts the soil with regard to phosphorus and potassium, but when nodulation is optimal, the crop leaves a very nitrogen-fertile soil for the following crop. Nodules develop only when Rhizobium japonicum is available in the soil. Rhizobia that may be effective in conjunction with other legumes may not fix any nitrogen with soybean. Potassium stimulates the action of the nodules. In many soils inoculation is necessary to obtain a fully developed crop. Inoculants are available either commercially or at university or government laboratories and can easily be applied to the seeds on the farm. Once established in a particular field, Rhizobium may remain available for many years, even without soybean cultivation. High nitrogen application reduces the effectiveness of nodulation. Spacing is 45–60cm between rows and 10–25cm between plants, depending on the bushiness of the plant (133 000–200 000 plants/ha). The seed rate is 40–60kg/ha.

Leaf-eating caterpillars, army-worms and bean fly may attack the young crop; pod-sucking stink bugs at later stages. Small damage by bugs may harm seed quality. When the damage is significant, yields may decline dramatically. The only solution seems to be insecticide sprays.

Major diseases include bacterial blight (*Pseudomonas glycinae*) and bacterial pustule (*Xanthomonas phaseoli*), the fungi pod and stem blight (*Diaporthe phaseolorum*), rust (*Phakopsora pachyrhizi*) and purple stain (*Cercospora kikuchii*), and the viruses soybean mosaic virus. Several of these diseases are seed borne.

Timing of harvesting is difficult, especially in hot dry weather. Shattering of seeds can cause significant losses. Harvesting small fields is best done quite early. Uprooted plants are heaped on a tarpaulin as soon as the foliage has dropped. Plants are left to dry, until threshing by beating with sticks becomes easy. Clearly, rain during this drying period may spoil the seeds.

Seed production

Selection
Being a self-fertilizing crop, variety maintenance selection is easy. Off-types are easily observed on plant type, colour of hairs and maturity period. Since seeds of different varieties are very similar, selection after harvesting is difficult. When a mixture has to be purified, a simple line selection will be very effective.

Seed-transmitted diseases
o Bacterial pustule: small yellow spots with raised brown pustules at the lower leaf surface, sometimes also on the stems.
o Bacterial blight: brown, angular spots forming dark-brown dead areas with yellow margins, spreading quickly in wet, windy weather.
o Pod and stem blight: sunken brown spots girdling stems, generally at the base of a branch or leaf stalk, forming black spores towards crop maturity. These diseases may survive in undecomposed crop residues. Control measure: use resistant varieties, crop rotation, destroy volunteer plants, and use seed from fields without symptoms where possible.
o Purple stain: clear purple specks to large blotches on the seeds. Control measure: seed selection
o Soybean mosaic virus: dark green mosaic, often with blistering of the leaf. Plants may become stunted. Control measure: roguing of plants with symptoms in the seed field.

Storage
Soybean seed is a notoriously poor storer. Delay of seed drying and slightly sub-optimal storage conditions seriously reduces seed viability. Under

tropical conditions the only solution is to shorten the storage period to the minimum by planting harvested seed within a month, even if it has to be in a dry season (under irrigation).

The problem may start in the field, i.e. when harvesting cannot be done under dry conditions, and when drying cannot be done immediately. When seed cannot be stored at less than 45 per cent relative humidity of the air, storage for 7–8 months is unsuccessful. When a soybean seed has to be stored during a humid season, the only way is air-tight packaging of well-dried seeds in e.g. a polythene-lined oil drum.

Options for improvement
Soybean seed is difficult to produce both for an individual farmer and a specialized seed enterprise. The main reasons are the diseases and in particular the poor seed storage under tropical conditions. The latter aspect is particularly difficult for large-scale local seed producers. In areas where seed can be produced in the off-season, e.g. in dry paddy fields when the main crop is grown in upland conditions, a very profitable seed business can emerge. Special attention can be given to improved local storage methods and disease control.

Sweet potato

Other names	Batata doulce, camote, kinkio, dantiu, anago, kumara	Market potential seed	poor
		Ease of seed production	easy
Propagation	vegetative		
Storage seed	very poor		

General description
Sweet potato (*Ipomoea batatas*) is a herbaceous plant that is usually grown as an annual crop. Carbohydrates are stored in the roots. There are varieties with long and short vines, that can be used for propagation, and for fodder and vegetable. Because of the limited storability of the roots and the possibility of maintaining sweet potato plants in the field for a long time, farmers harvest tubers and cuttings according to their immediate needs.

Agronomic requirements
Sweet potato is a crop of the warm humid tropics and grows well with 1000mm rainfall. It is, however, reasonably drought resistant because of its capacity for regrowth after defoliation. Temperatures below 12 °C lead to leaf damage and yield reduction. Edible roots develop best after the vines have stopped growing. Depending on the variety and the growing conditions, roots are harvested from 70 to 360 days after planting. The roots

suffer from water-logging, so well-drained soils are necessary, or planting on hills or ridges to avoid rotting. Sweet potato reacts positively to added nitrogen and has a relatively high potassium requirement.

Notes on multiplication

Selection
Since sweet potato is vegetatively propagated, the offspring of one plant are genetically identical. Local varieties may be a mixture of clones. Especially in Central America, the Pacific and South-east Asia, genetic diversity can be found with respect to leaf and root shape, skin and flesh colour and sugar content. The International Potato Centre (CIP) maintains a large collection of clones.

Production of vines
Multiplication does not interfere with crop production, since these activities concern different plant parts. Cuttings are generally the top part (30–40cm) of the vines, but lower parts and slips or sprouts may also be used for planting. Roots develop easily from nodes on the cuttings which are planted approximately 20cm deep.

Botanical seeds can be used for propagating sweet potato in breeding programmes, but in common multiplication this is rare.

Diseases
Sweet potato suffers from virus infections which are transmitted by insects. The symptoms include mosaic patterns and malformation of the roots. Bacterial and fungal diseases may cause some problems, but the most serious enemy of sweet potato is the sweet potato weevil. It lays eggs in stems and roots and the larvae eat their way through the plant. Removal of host weeds (other Ipomoea species), frequent hilling and the avoidance of soil cracks prevent the access of the weevil to the roots where it does most damage.

Selecting vines that do not carry obvious disease symptoms can significantly increase the health of subsequent crops.

Options for improvement
The production of planting materials commonly goes hand in hand with crop production. Because of the poor storage of the vines, large-scale production of propagating materials is very difficult, which may be a serious problem in the large-scale diffusion of new varieties. Plant breeding concentrates on the development of high-yielding varieties with a high consumer acceptability.

Sunflower

Other names	*tournesol, girasol, mirasol*	Market potential seed	fair
Propagation	cross-fertilized by insects	Ease of seed production	easy
Storage seed	good		

General description

Sunflower (*Helianthus annuus*) is quite a new crop to Africa and Asia. It has not become a common household processed oil, and is thus mainly grown as a cash crop for local oil mills. Sunflower varieties range in height from 60cm to 5m; head diameter from 10 to 50cm. The crop is adapted from the equator to temperate latitudes, but not to the humid conditions of the tropical lowland, where diseases take their toll (Botrytis). The plant is cross-fertilized by bees and other insects.

Agronomic requirements

Sunflower has a very well-developed root system, which can extract minerals that other crops are unable to reach. Sunflower is particularly sensitive to boron deficiency. Row planting is done with a row spacing of 60–90cm and 30cm between the plants in the row (population 40–55 000 plants per ha). Dwarf types can be planted much closer. Sowing should be done in a well-prepared moist soil and should be timed such that harvesting can be done in a guaranteed dry and warm period. When the crop is established it can tolerate drought and heat very well.

A sunflower crop will not easily cover the soil. Weeds can therefore develop into a serious problem for both the sunflower and the subsequent crop. The heads are harvested when the outer bracts turn yellow and brown. This secures the largest percentage of mature seeds in the head, which matures towards the centre. Heads are dried until the seeds come loose. Important diseases are the cool weather disease grey mould (*Botrytis cinerea*), rust (*Puccinia helianthi*) and stem rot (*Sclerotinia sclerotiorum*). The most important pest in most sunflower growing are birds, such as Quelea. They can very effectively destroy a crop only days before harvesting.

Seed production

Selection

Sunflower is an insect-pollinated crop. This means that, especially in small sunflower fields, pollen may be carried over large distances. The risks of contamination of the variety with foreign pollen is considerable. Isolation requirements may be as much as three kilometres, which is not easily maintained. Sunflower for seed should therefore not be planted in very small seed production fields, but be part of normal sunflower plantings. The best way to maintain the genetic purity of the variety is to plant seed of selected heads in the centre of a field planted with seed of the same origin. This field centre should then be harvested separately for seed. Selection criteria are single heads (as opposed to split-heads), uniform maturity (to allow for early harvesting and to reduce bird damage), and uniform-coloured seeds. When for such major characteristics a variation exists within a variety, serious selection with the assistance of trained breeders may be warranted.

Production and storage
With carefully selected sunflowers in a seed production programme, losses due to birds have to be avoided. Racks can be built near the homestead to dry the harvested heads. In larger-scale seed production (more than 0.5 ha), drying of the heads can be accelerated by cutting them and pinning the cut heads back on to the stem. The detached heads dry quicker and the reduced height allows bird scarers to do their work more effectively.
A major problem with sunflower is that good seed and underdeveloped seeds (empty hulls) are hard to distinguish with the eye. There is thus a risk of carefully storing a bulk of seed with low germination capacity. The result may then be that insufficient good seed is being stored. Proper inspection by squeezing a number of seeds is therefore necessary.

Options for improvement
Sunflower is in most countries a commercial crop. Even though small hand-operated presses are available, most smallholder and large-scale sunflower producers depend on oil mills to process their produce. The oil mills often supply seed of varieties with high oil content and good pressing characteristics (white silica in the seed coat is an abrasive in the mills, so the owners prefer the black-seeded varieties). In many cases this seed is just taken from selected lots of produce; sometimes it is hybrid seed from formal seed producers. Small-scale commercial production of seed is therefore limited in scope, unless the operation can be connected to a mill. Improvement may concentrate on the proper maintenance selection of sunflower varieties, since the risk of genetic degeneration is high.

Wheat

Other names	blé, froment, trigo	Market potential seed	good
Propagation	self-fertilizing	Ease of seed	
Storage seed	poor	production	fair

General description
Many species and sub-species of *Triticum* exist, of which bread wheat (*Triticum aestivum v. aestivum* L.) and durum wheat (*T. turgidum* L. var.

252

durum (Desf.)) are the most important ones. Durum wheat has a higher protein content and can be used for pasta preparation. Both wheats are essentially annual grasses adapted to an arid climate with a rainfall season of limited duration, and were domesticated in the Near East. Bread wheat has a greater drought tolerance. Although wheat can tolerate modest levels of drought reasonably well, good water supply enhances yield; moisture requirement varies between 300 and 900mm. Wheat is an erect plant with a height of up to 1.5m in the case of landraces, and a height of 1m or shorter in the case of modern varieties. It tillers profusely if plant density is low and growing conditions are favourable. A terminal ear of 5–10cm length contains rows of kernels that are easily threshed and form the basis of a multitude of foods.

Agronomic requirements

Wheat is mainly grown in temperate climates; production in lowland tropical environments is not possible. In mountainous areas (Mexico, Central Africa) and in countries with a winter season (Bangladesh, India, Zimbabwe) it is possible to grow wheat of reasonable to good quality, especially when the weather is dry during harvest. Production in the dry season under irrigation is an alternative for such countries, because of the relatively low disease pressure (e.g. in Zambia) and because of the certainty of adequate water supply during the grain-filling stage. High radiation levels improve yield.

A seed rate of 80–150 kg/ha is recommended, with a spacing of 20cm between rows. Seeds are sown 2–4cm deep. Under fertile conditions one can use as little as 30 kg/ha for seed production (for quick bulking of a new variety). Under such conditions the highest seed multiplication rate can be obtained, although yield will be sub-optimal. Landraces, which are usually tall and have a weak stem, lodge easily (especially when nitrogen is applied).

Storability for the seed is poor, especially during the humid wet summer seasons.

Seed production

Selection

Because the crop is self-fertilized, a distance of one or a few metres between fields of different varieties is sufficient. This distance is advisable to avoid mechanical mixing during sowing or harvesting. Rotation with a non-cereal crop is advised in order to avoid the build up of diseases. An interval of two years is usually sufficient. Some grassy weed seeds may be difficult to remove once the seed is harvested. Therefore it is important to keep the field and its edges free from weeds.

Seed production is best done on heavier soils with good water-holding capacity and good surface drainage. Contamination with weeds or other crops should be avoided. Fields that have been used in the previous season for other small and large grains should be not be used, to avoid seed contamination and diseases. Sowing should be done at sufficiently high density to suppress the growth of emerging weeds.

Wheat varieties are commonly quite uniform and the fertilization method allows for relatively easy selection when necessary. Taller off-types are normally easily recognized, but shorter ones often remain unnoticed. Latest advances in breeding have produced hybrid wheats, which are only suitable for farming systems in which seed can be purchased annually. A large germplasm collection is maintained by the International Maize and Wheat Improvement Center (CIMMYT) in Mexico. Many national germplasm banks also possess collections.

Seed-transmitted diseases

The main seed-transmitted diseases are fungal infections such as loose smut (*Ustilago tritici*) and common bunt (*Tilletia caries* and *T. foetida*). These diseases are effectively prevented by adequate seed dressing. If present in the field, all plants with symptoms have to be removed immediately before they become infectious when their kernels break. If more than 0.01 per cent of the heads is infected, the seed should ideally not be used to produce another seed crop. Therefore, very rigid disease control is needed in seed crops.

The gall nematode, *Anguina tritici*, is dispersed by seed galls which are mixed with grain, however, larvae may also be carried by seed of relatively normal appearance.

Partly discoloured brown or blackish seeds are often seeds infected with carnal bunt (*Tilletia tritici* or *T. indica*); seeds spots may be caused by *Pseudomonas* or rusts. Seeds may also be entirely transformed into fungal bodies. Discoloured seeds should thus be removed from a seed lot. Nematode galls (ergot) look like seeds, but are smaller. Seeds with scab infection (*Fusarium* spp., such as *F. graminearum*) can show pink discolourations and are often smaller and shrivelled, and are called 'tombstones'. Farmers should try to hand-pick their own seed systematically. In large-scale production, however, this is not possible.

Processing and storage

The threshing of wheat seed should be done when the seed moisture content is below 19 per cent. Since the germination of wheat seed drops quickly relative to most other cereals, the seed should ideally be stored at a moisture level below 12 per cent (in air-tight containers even down to 9 per cent).

Common weeds are several species of *Brassicaceae* such as *Raphanus, Sinapis, Rapistrum*, and *Convolvulaceae* (*Convolvulus*). Difficult to remove are wild oats (*Avena fatua* and related species) and admixtures of other cereals. Processing the seed with an air-screen cleaner is thus very useful: it is one of the most effective tools for reducing the amount of weeds in the field. Depending on the variety, a top screen with round perforations is needed to remove straw and other large debris. The good seed must readily fall through this screen before it reaches halfway down. Usually this is achieved with holes of about 5mm diameter. Next, the seed must fall on to a screen with oblong perforations to remove weed seeds slightly thicker than the good wheat grains. It is important to select this screen carefully: too narrow will result in excessive seed losses, too wide will have no effect. Finally, the grain must be fed over a bottom screen with round perforations to remove sand, undersized wheat and weeds. In a well-equipped processing plant an indented cylinder separator will be used to remove broken grains and round-seeded weeds.

Wheat seed stores relatively well when dried.

Options for improvement

Wheat seed can easily be produced by competent farmers in most wheat-growing areas. The commercialization of wheat seed production is therefore rather difficult, except in areas with winter-grown wheat and a hot moist summer. Because wheat seed does not store well in such conditions, there may be an effective demand for good wheat seed. Also, in marginal wheat production areas, the build-up of seed-transmitted diseases may be a good reason for farmers to look for seed elsewhere. In the first case, it may be best to produce wheat seed in a region with less humidity during summer, and transport it to the wheat grain-producing region. Alternatively, when seed can be dried well after harvesting, seed could be stored in air-tight sacks or silos (or polythene-lined oil drums with some charcoal as a drying agent).

In the case that seed-transmitted diseases are a serious problem, seed dressing with Captam or Thiram will provide a good service to the farmers. Improving wheat varieties basically involves the selection of better varieties or the purification of the existing ones. Most breeding efforts concentrate on yield level and stability, and improvement of tolerance to pests and diseases. Reduction of height in the case of landraces can be attempted, which is likely to result in reduced lodging and increased yields. End-of-season drought may be a reason to select earlier maturing varieties. The grain quality (e.g. bread-baking quality) is usually an important selection criterion, and can easily be tested by producing the desired food product.

VEGETABLES

There are many plant species used as vegetables. Many of the species important for their use as vegetables are divided into groups or sub-species. Sub-species may be divided into variety groups. It would be beyond the scope of this publication to describe them individually. In South-East Asia alone, 84 major vegetables and 128 minor vegetable species are grown (Siemonsma and Piluek, 1994 *Plant Resources of South-East Asia 8: Vegetables*, Wageningen, Pudoc, and Bogor, Indonesia, PROSEA). For this reason, only the main groups of vegetables are described: Alliums, Cucurbits, Crucifers, Solanaceous, and tropical leafy vegetables.

Alliums

Crops	Onion, shallot, garlic, bunching onion, leek	Market potential seed	good
Propagation	cross-fertilizing	Ease of seed production	difficult
Storage seed	poor		

General description
The family of the alliums includes both vegetatively and seed propagated species. Alliums produced from seed belong to the biennials, and the flowers are produced only after a vernalization (cold period) of the vegetative plant. This poses serious problems for seed production under warm tropical conditions. In the tropics, onions are commonly produced from imported seed or seed produced in cool tropical highlands. Alternatively, related crops (shallot, garlic) are produced that can be vegetatively multiplied using dry sets. These are small bulbs that can regrow and develop a new plant. Alliums are produced for their more or less pungent leafy organs, such as the bulb or the (lower) leaves. A wide variety exists in pungency, which is a major distinguishing characteristic of different varieties. Many alliums are day-length sensitive, i.e. bulb formation occurs during 16 hours' light (temperate varieties) or during 12 hours' light (tropical varieties) per day. Introducing new varieties into a particular region has to take this day-length sensitivity into account. Bolting occurs after vernalization. The requirements vary strongly among varieties.
Flowers are cross-pollinated by insects such as flies and bees, but some selfing may occur.

Agronomic requirements
Allium crops can be grown under a wide range of climatic conditions, but they succeed best in a climate without excessive rainfall. Dry conditions are necessary during the maturation of the bulbs. Seeds can be sown in nurseries or directly into the field. Dry sets are planted into the field directly.

256

Under dry conditions, young plants need well-planned irrigation. The crops are quite sensitive to damage from to mechanical weeding. Potassium is an important nutritive factor, producing high-quality seed even when given as fertilizer during the bulb-production season and not during the flowering season.

Seed production
Production of planting material of vegetatively propagated species follows common crop production procedures, including roguing of unwanted types (plant architecture mainly) and diseased plants. Production of true seed of e.g. onion requires very specific conditions. The process can be divided into three periods: bulb production, vernalization and seed production.

Bulb production requires quite dry and it is possible in hot conditions, the vernalization period has to be cold, and the flower and seed production period has to be moist at the start and dry at the end. Very good conditions for this sequence occur in the area of origin of the onion (Iran-Afghanistan). Ecologically similar areas can be found in the USA and India.

In tropical areas, dry conditions may be found in the lowlands for bulb production, but cool conditions in the highlands are often too humid (clouds and rain) to guarantee good seed ripening. An alternative is vernalization of harvested bulbs in a cold store (an ice or a fish factory), after which the bulbs can be replanted in the warm and dry conditions. In a cross-fertilized population of onions, differences in vernalization requirement are likely. It is thus possible to select for plants that bolt under warm conditions. This offers the potential to produce seed under ambient conditions, but at the same time this selection for early bolting also causes unwanted flowering during vegetable production.

Selection
Selection in onion varieties concentrates on the bulb characteristics, such as size, shape and colour. 'Thick necks' (split bulbs) and early bolting plants are discarded. In leeks, leaf colour and stature of the plant are major selection criteria.

A strict selection against disease symptoms on leaves and bulbs can significantly reduce disease levels in the seed, and increase seed yields. Being a cross-fertilized species there is commonly a wide variation within local varieties, which allows for advances by selection.

In vegetative propagation the main issue is the avoidance of transmitting the many diseases that attack the bulbs. Healthy plants have to be selected. Providing a clean stock of plants (through specialized tissue culture techniques) can be a great help to farmers with disease-infested allium crops.

Seed-transmitted diseases
The main seed-transmitted diseases are Alternaria (purple blotch), Botrytis (damping off), Colletotrichum (smudge), Peronospora (downy

mildew), Erwinia (soft rot), and onion yellow dwarf virus. Seed treatment (Thiram/Benomyl) can be a very effective way of reducing disease incidence, but against some diseases careful handling and long rotation (Erwinia) are necessary.

Harvesting and cleaning
The timing of harvesting seeds is quite difficult. Onion and leek seed are best harvested when 5 per cent of the capsules on individual heads are shedding ripe seeds. Harvesting later results in significant seed yield losses and earlier harvesting gives a high percentage of immature seeds which reduces the germination percentage. Even where heads are hand-picked, some seed yield losses have to be accepted. Quick drying of the heads and seeds is necessary since allium seeds are very sensitive to high moisture content. Dry seed will accumulate moisture under high air humidity, so in humid tropical conditions dry seed has to be kept in moisture-proof containers. Expensive laminated polythene packs are useful, but also soda bottles closed with a cork and some candle wax may serve the same purpose.

Options for improvement
Onion and leek seed are commonly available in tins or small seed packets. Local seed production could help farmers obtain a good additional income. The main problem is, however, that high-quality seed can be produced only in particular areas. Such areas have to have well-drained soils, a hot dry climate in combination with reliable irrigation for both bulb and flower production, and a cool season for vernalization. In addition, sufficient insects have to be present during flowering for pollination. Farmers in such areas could develop a profitable business if a good variety can be identified and when disease epidemics can be avoided.

Industrial development can, however, open up cost-effective alternatives for vernalization and production possibilities in formerly unadapted regions. Fish-processing plants may allow an onion-seed producer to use a cold room for vernalization of the bulbs. A simple trial will show the vernalization period for the available varieties.

For all other farmers, variety trials under local crop management practices can yield interesting new materials. International plant breeding has produced varieties that are adapted to different climatic conditions and that are resistant to various diseases. Many commercially available varieties are, however, hybrids which cannot be multiplied locally.

Vegetatively propagated alliums are commonly multiplied by small-scale farmers for their own use. Improvement of local production can be attained by eliminating diseases and possibly improving the drying process of the dry sets. Where diseases are present in virtually all vegetative shallot or garlic materials in a particular region, commercial production of disease-free planting materials can be very profitable.

Cruciferous vegetables

Crops	cabbage, turnip, mustard, radish	Market potential seed	good
Propagation	cross-fertilizing	Ease of seed production	difficult
Storage seed	poor		

General description
Cruciferous vegetables include a very variable group of crops. Brassica is an important genus within the family of Cruciferae, and is divided into a number of species, of which Brassica oleracea is one. Others are Brassica napus (beetroot), Brassica nigra (mustard) and Brassica carinata (Ethiopian mustard). Some of these crops are grown for their leaves (cabbages), others for their stems (knolkohl), small side-shoots (brussels sprout) or roots (turnip, radish). These widely varying uses require very different agronomic practices, such as soil requirements, spacing of plants, etc. Flowering of the various crops within this group depends to a greater or lesser extent on a low temperature vernalization. This means that the vegetative crop has to remain in the field during a cool period. This may be the winter in temperate climates or can occur at higher altitudes in equatorial areas.

Cruciferous crops are cross-fertilized by various insects – bees and flies being the most effective. Different crops in the Brassica oleracea species readily cross (white cabbage, cauliflower, kohlrabi, etc.) and also crosses between cruciferous species are possible.

Agronomic requirements

Seed production
Since cruciferous vegetables are grown mainly for their leaves (cabbage) or roots (radish), seed production is a specialized operation. Plants have to be reserved for seed production beyond the vegetable harvesting stage. In warm conditions, seed cannot be produced at all because the plants do not flower. Cabbage, for example, has become an important vegetable throughout many tropical areas even though local seed production is impossible. In such areas cultivation is based on imported seed, i.e. seed imported by seed companies or the government, or seed provided by local merchants from mountainous areas in the country.

Apart from the cold requirements, some special techniques may be necessary to obtain seed. Mature cabbage heads need to be cut cross-wise to allow the flower stems to emerge. This has to be done after checking the trueness to type. Most varieties that are produced by commercial seed companies are F1 hybrids. Producing seed from such varieties is in some cases impossible (incompatibility or sterility), and in most cases results in frustratingly poor-quality offspring.

Selection

The main selection criteria in cabbages are a lack of branching, and head characteristics. The main characteristics to select for in turnip and radish are the root shape and colour. In all cruciferous crops a major criterion is the presence of early bolting plants. This is very important since flowering plants have no value in the vegetable market. By selecting flowering plants under warm conditions there is thus an added risk of selecting for early bolters. All cruciferous crops are predominantly cross-fertilized and open-pollinated varieties are relatively heterogeneous. This large variation between plants can result in the reduction of head or root quality when varieties are not properly maintained. Variety maintenance includes the strict selection of a number of plants and allowing the selected plants to flower together. In order to avoid inbreeding, at least 100 but preferably 200 plants should be selected. With root crops (radish, turnip), the plants can be uprooted, selected for their root size, shape and colour, and replanted in a seed production block. With cabbage plants, selected plants can also be replanted after careful uprooting, but more often, selected plants remain in the field and off-type plants are removed. Because of the insect pollination, relatively large isolation distances of one to five kilometres are required by official certification systems.

Seed-transmitted diseases

Cruciferous species are attacked by a wide range of seed-transmitted diseases. Alternaria (grey leaf spot, black spot), Ascochyta (leaf spot), Colletotrichum (Anthracnose), Leptosphaeria (black leg) and the seed- and soil-borne Sclerotinia (white blight) and Plasmodiophora (clubroot) are the most common fungi. The bacterial diseases Pseudomonas (bacterial leaf spot) and Xanthomonas (black rot) can also be seed transmitted. Long rotation cycles are the only feasible way to reduce the latter two diseases. In rotation it is very important to remove host plants (cruciferous weeds and mustard and rapeseed plants) away from the edges of the field. The diseases are also transmitted through water, so in hilly areas the high-lying fields also have to be be kept free of these diseases. Chemical soil disinfection with e.g. methylbromide is theoretically possible but expensive, dangerous and hardly ever fully effective. Particularly where special areas are designated for seed production (e.g. high-altitude areas) special care has to be taken to avoid the introduction of such seed- and soil-borne diseases (quarantine measures).

Harvesting and cleaning

Inflorescences of cruciferous crops are widely branched racemes. Flowering starts from the centre and develops to the outer branches with time. Seed ripening follows the same path. This means that the harvesting time is crucial. Early harvesting results in a large proportion of unripe seeds that will not germinate. Late harvesting results in significant seed yield losses due to shattering. Flowering stalks can be harvested as soon as the colour of the fruit changes and when the seeds of first fruits have shattered.

Post-harvest operations are important: threshing well-dried inflorescences is relatively easy, but harvesting during wet weather renders threshing impossible and introduces mould growth on pods and seeds. Careful seed cleaning can remove unripe seed to some extent, especially with a gravity table, but very advanced machines are required to upgrade early harvested seeds.

Seeds of cruciferous crops are very sensitive to moisture. Keeping seeds under ambient conditions in a humid climate commonly results in disappointing germination. Drying the seed directly after harvesting, and subsequent packaging in moisture-proof containers is necessary for successful seed storage.

Options for improvement
Imported cruciferous seeds are commonly expensive. Local production of seed could assist farmers to obtain a good additional income. The main problem is, however, that high-quality seed can be produced in specific areas only. Such areas have to have well-drained deep soils, a cool season for vernalization and guaranteed dry weather during the harvesting season. Furthermore, such areas should not be infested with wild relatives of the cultivated crops, since outcrossing beyond the species barrier are common. Farmers in such areas could develop a profitable business if a good variety can be identified and when disease epidemics can be avoided.

For all other farmers, variety trials under local crop management practices can yield interesting new materials. International plant breeding has produced heat-tolerant varieties that are adapted outside the original cultivation areas.

Cucurbits

Crops	Water melon (Citrullus), melon and cucumber (Cucumis), pumpkin and marrow (Cucurbita), bitter gourd (Momordica), Loofah (Luffa), etc.	Market potential seed	fair
		Ease of seed production	rather easy
Propagation	cross-fertilizing		
Storage seed	good		

General description
Cucurbits are annual vine crops that produce fruits that are sometimes considered vegetables (bitter gourd), and sometimes fruits (melons). The

majority of species in this group have unisexual flowers borne on monoecious plants, i.e. each plant carries both male and female flowers. Cross-fertilization through insect pollination is the rule, but selfing can occur in some species.

Agronomic requirements

Cucurbits are normally produced by direct seeding. Seeds germinate slowly, especially during cool weather, and weeding during the early stages of the crop is vital. Some cucurbits can be grown on trellises (cucumber, gourds), whereas others are commonly grown on the flat (melons, water melon). Seeds are planted on ridges or between ridges depending on rainfall patterns and the availability of irrigation. Irrigation is required under many conditions. Care has to be taken to avoid a humid microclimate for the leaves and the fruit, since this readily triggers the development of diseases.

Seed production

Selection

Varieties, and even different species within this group, are capable of crossing, and large isolation distances are required to avoid degeneration of varieties. Isolation of up to three kilometres are used in official certification and depends on the feeding range of the insects (mainly bees) that pollinate the crop. Selection in all species can be done on the immature fruits. Selection on the basis of the characteristics of the first fruits allows for an early roguing of off-type plants. When a wrong-type fruit is discovered, care has to be taken to remove the whole plant. This can be difficult with crops of which the vines are intertwined.

Cucurbits are cross-fertilized, which means that the variety may degenerate when not maintained.

Seed-transmitted diseases

Common diseases of cucurbits are seed transmitted. These include, for example, anthracnose (Colletotrichum), Alternaria leaf spots, Fusarium wilt, Didimella (black rot), mosaic virus and Pseudomonas seed rot. In several cases it is difficult to control these diseases without chemical sprays.

Harvesting and cleaning

Fruits destined for seed production have to be left in the field beyond the vegetable harvesting stage. In some crops, discoloration of the fruits indicate ripe seeds (cucumber); in others, such as watermelon, ripeness is shown by the tendrils withering, or can be assessed by tapping the fruit: a hollow sound indicates ripeness. Commonly, the rind of ripe fruits has hardened. Since the vines continue to grow after the first fruits develop, harvesting of individual fruits cannot be done in one round. Under very dry conditions, however, ripe fruits may be left in the field until all fruits are ripe. When fruits are picked at the vegetable stage, the plant will continue to grow longer than when the first fruits are left on the plant for seed. This means that a special planting for seed production can be done at a closer spacing.

Seed extraction is a special operation. Where seeds are located in the central axis of the fruit (cucumber), this part can be scooped out. In other crops, such as water melon, the whole fruit may have to be macerated to free the seeds. The seeds have to be washed in running water to separate the seed from the pulp and to remove attached 'slimy' fruit materials from the seed. Mature seeds will sink, while pulp and immature seeds will float. In some crops, fermentation can remove the last bits of pulp, but this can cause discoloration and some reduction in germination. Extensive washing is better.

Seed has to be dried as quickly as possible after washing. There is a risk of the seeds sticking together due to tiny bits of pulp remaining on the seed. Carefully rubbing the seed during the drying process will avoid the formation of clumps of seed.

Options for improvement
The size and shape of cucurbit fruits may vary considerably. This variation is commonly not an important characteristic. A strict selection within a population of cucumbers, pumpkins or marrows in order to make the variety more uniform commonly serves no purpose. Only where the market demands a uniform product will selection be useful, and seed improvement activities may have a role. This may involve selection within an existing variety or by multiplying a well-tested introduced variety. Since cucurbits are cross-fertilizers, care must be taken to avoid in-breeding. A seed crop should preferably be grown from a mixture of 200 selected fruits (from different plants). Introduced varieties of cucumber, water melon and other cucurbits are often F1-hybrids, however, which generally do not produce a uniform offspring when multiplied. In addition, since well-dried seeds store quite well, and since each fruit produces large quantities of seed, little market potential can be expected for cucurbit seeds.

Solanaceous vegetables

Crops	Tomato, sweet pepper, aubergine	Market potential seed	fair
Propagation	self-fertilizing	Ease of seed	quite
Storage seed	good	production	easy

General description
Solanaceous vegetables are among the most common vegetables worldwide. Tomato and sweet pepper originate from Latin America and aubergine (eggplant) from Asia, but these vegetables have become part of

the diet in most areas of the world. Solanaceous vegetables are (mainly) self-fertilized. Varieties have either a determinate (bush) or an indeterminate (trailing) plant habit. Solanaceous vegetables are grown as annual crops, but under particular growing conditions they can be perennial plants. Each fruit contains a large number of seeds.

Agronomic requirements
Seedlings are commonly produced in seed beds and transplanted into the field. Special care has to be taken to harden the seedlings before transplanting. Soils have to be fertile and well drained to produce a good vegetable crop. The crops in this group share a number of common diseases, and rotation therefore has to exclude other solanaceous crops, including potato. Earthing up of the plants may result in the formation of adventitious roots which will increase the sturdiness of the plant and provide more moisture and nutrients to the plant.

Seed production
Seed production agronomy follows normal crop production methods to a large extent. The spacing of the plants may be less than is common in crop production. During crop production, the first fruits are removed at the marketable stage, after which the plant can support further vegetative growth. When the fruits remain on the plant until full ripeness of the seed, however, vegetative growth is reduced, which reduces the total plant size and thus allows for the possibility of more plants per hectare. It may be economically interesting to pick the first fruits for consumption and leave the last fruits for seed, but this is not a good practice, since seed-borne diseases may have more time to infect the plant and the fruit, and carry the disease forward to the following seasons.

Selection
The selection of solanaceous vegetables is very straightforward. Being self-fertilized, few off-types occur, and important characteristics are readily visible as soon as the first fruits develop. In sweet pepper and eggplant, however, some outcrossing may occur, which can be a nuisance when uniform fruit types are required in the market, or a blessing where the genetic diversity offers opportunities for improvement.

Many commercial varieties are F1 hybrids which do not reproduce the variety when multiplied. It may, however, be possible to select outstanding plants from the segregating population produced by planting the seed produced on such a hybrid plant.

Seed-transmitted diseases
Various important diseases are seed transmitted. Examples are the fungus diseases Alternaria (early blight), Colletotrichum (Anthracnose),

Fusarium, Rhizoctonia and Verticillium (wilt), Corynebacterium (canker), and mosaic and bunchy top viruses. Bacterial wilt (Pseudomonas) is a major bacterial disease that is not, however, seed transmitted. A general guideline is that fruits should be harvested for seed only if the plant does not show symptoms of these diseases. Treatment of the seed with Captan greatly reduces seedling diseases (e.g. damping off).

Harvesting and cleaning
Fully ripe fruits can be picked for seed extraction. Fleshy fruits (tomato) are crushed and left to ferment for a day (or two or three depending on the temperature). This process removes the mucilage surrounding the seeds. Too long a fermentation, however, strongly affects seed quality. Alternative methods include the use of sodium carbonate or even hydrochloric acid. The latter method is definitely not suitable for home use even though this method is reported to reduce possible virus infection. After fermentation, the seeds are washed and dried, during which time they have to be brushed from time to time to avoid clogging. Eggplant and sweet pepper can be extracted by the same wet method or by a dry treatment. Dried eggplant fruits are beaten, during which the seeds separate from the surrounding material; sweet pepper seed can be removed with a knife.
    Properly dried seeds store relatively well.

Options for improvement
A wide range of varieties are commonly produced locally. Collection of different types in the area and from neighbouring communities may yield many different types that can be selected from. International plant breeding also produces many new varieties every year that can be included in local trials as well. These may be particularly interesting for the introduction of disease resistance. Most of these international varieties are hybrids, however, which may not retain important characteristics after local multiplication.
    The market opportunities for seed of solanaceous vegetables is in most areas not very large. New varieties may spread quickly since many self-fertilized seeds can be extracted from any (marketable) fruit. Significant improvements through seed can, however, be obtained in areas with high (virus) disease incidence. Knowledge about the identification of diseases, proper control and avoidance of transmission (early planting, harvesting of the first fruits from symptomless plants only) can greatly reduce disease transmission and produce better seed, which subsequently produces more marketable fruits.

Local leafy vegetables

| Crops | Amaranth (Amaranthus spp), Indian spinach (Basella spp), *soo* (Celosa spp), water spinach (Ipomoea spp) | Propagation | various |
| | | Market potential | limited |

General description
A wide range of plant species are used as leafy vegetables. The crops mentioned in the heading of this section include only the most widely distributed ones in the world next to lettuce (Lactuca), endive (Cichorium), spinach (Spinacea) and cabbages (Brassica), which originate in temperate climatic zones. Leaves are also collected from crops that have another primary product, e.g. sweet potato, okra, jute, and from natural or weedy populations, e.g. amaranth weeds in field crops. Also, field crops such as cowpea and cassava produce edible leaves as an important by-product, and should be considered local vegetables as well.

All local leafy vegetables have one thing in common – the produce is poorly commercialized because of the difficulty of storing the easily withering leaves. This in turn results in relatively little investment both in terms of local technology development and formal research, and a very limited commercial seed supply.

Leafy vegetables are, on the other hand, very nutritious and important components in all diets worldwide.

Agronomic requirements
Leafy vegetables are commonly grown in small gardens or in complex intercropping systems. In intercropping, the agronomy has to comply with the main constituents of the mixtures, which means that the minor components simply have to adapt to the main crops. In backyard gardens, however, vegetables can get specific care and attention. Leafy vegetables generally require a soil that is rich in organic matter and especially available nitrogen and potash. Some crops have a particularly high uptake of micro-nutrients such as magnesium (spinach).

Because of their vigorous vegetative growth, most leafy vegetable crops are rarely affected by weeds, and most are quite susceptible to drought. Insects and diseases can cause serious damage to the commercial product. This problem is aggravated by the fact that the commercial product (the leaf) is at the same time the sprayed product in chemical control. This means that there is a serious risk of chemical residue consumption when these crops are chemically protected. This is, of course, less relevant when producing seeds.

Options for improvement

Relatively little attention has been paid to the formal breeding of local leafy vegetables. Exceptions are amaranthus and water spinach in Asia, but in these cases breeding mainly involves selection among landraces. The lack of 'formal' interest may be a disadvantage for improvement, but at the same time this presents an opportunity to build on existing local knowledge. Selection within available populations is a good start for improving the crop for specified characteristics.

Since the product is the leaf, the quality of the product is easily spoilt by diseases. Avoiding seed-transmitted diseases can thus be very important. In may cases, however, local selection has developed very healthy leafy vegetable crops. In case a source of resistance is not available within the local populations, introduced materials may be tested. The Asian Vegetable Research and Development Centre (AVRDC) in Taiwan may have either materials or information on genebank collections. The FAO global information system on plant genetic resources and IPGRI may also provide such information. For example, Amaranthus collections are kept in Kutztown, Pennsylvania (USA), AVRDC and national centres in India (NBPGR) and Indonesia (LEHRI); and Ipomoea collections are kept at Kasetsart University (Thailand) and AVRDC.

For all seed production and vegetative propagation, local techniques have to be the basis for improvement. The little attention paid by formal researchers and seed producers may indicate that there is no market for seeds of local leafy vegetables. This may be true for large-scale seed production. Some seed companies in Asia have started to offer seed of the main leafy vegetable crops. This seed is mainly used for market production near large urban areas. There is, however, a possibility that specialized farmers who develop a superior selection from a local landrace may find a ready market for their seed.

Table D.1: Basic data for seed production (based on monocrop systems)

Crop	1000-seed weight (g)[1]	Planting rate (kg/ha)	Multiplication factor[1]	Growing period (days)	Isolation (m)	Seed storage[3]	Mating system
CEREALS							
Barley (*Hordeum vulgare*)	50	120	12	100–210	30	fair	self
Finger Millet (*Eleusine coracana*)	3	15	50	60–90	5	good	self
Maize (*Zea mais*)	300	25	80	90–150	300	fair	cross (wind)
Pearl Millet (*Pennisetum typhoides*)	7	8	120	60–90	300	fair	cross (wind)
Rice (*Oryza sativa*)	25	20/100[2]	20/80[2]	90–210	5	fair	self
Sorghum (*Sorghum bicolor*)	20	10	100	90–210	100	good	semi-cross (wind)
Wheat (*Triticum aestivum*)	40	120	12	120–210	5	poor	self
PULSES							
Bean (*Phaseolus vulgaris*)	200–600	60–120	10	80–120	5	good	self
Bambara groundnut (*Voandzeia subterranea*)	500–750	40	12	80–110	100	good	semi-cross (insects)
Broad bean (*Vicia faba*)	500	60	15	90–180	100	very good	semi-cross (insects)
Chick pea (*Cicer arietinum*)	500	45	20	120–180	5	good	self
Cowpea (*Vigna unguiculata*)	200	20	50	75–120	100	good	semi-cross (insects)

Lentil (*Lens esculenta*)	300	10	70	90–150	5	good	self
Pigeon pea (*Cajanus cajan*)	80	10	80	90–250	100	very good	semi-cross (insects)
Groundnut (*Arachis hypogaea*)	300	80	8	90–150	5	poor	self
OIL CROPS							
Sesame (*Sesamum indicum*)	3	6	120	60–90	300	good	semi-cross (insects)
Soybean (*Glycine max*)	150	50	20	80–160	5	very poor	self
Sunflower (*Helianthus annuus*)	40	10	80	90–140	600	fair	cross (insects)
VEGETABLES							
Beans (*Phaseolus vulgaris*)	400	80	10/20	90–150	5	very good	self
Cabbage (*Brassica oleracea*)	3	0.5	1000	450[3]	600	good	cross (insects)
Carrot (*Daucus carota*)	2	14	15	420[3]	600	fair	cross (insects)
Cucumber (*Cucumis sativus*)	40	2	100	45–90	600	good	cross (insects)
Eggplant (*Solanum melongena*)	3	0.12	1000	45–90	100	good	semi-cross (insects)
Bitter gourd (*Momordica charantia*)	250	2	100	45–90	600	fair	cross (insects)

Crop	1000-seed weight(g) [1]	Planting rate (kg/ha)	Multiplication factor [1]	Growing period (days)	Isolation (m)	Seed storage [3]	Mating system
Melons (*Cucumis melo*)	150	2	120	60–120	600	good	semi-cross (insects)
Okra (*Abelmoschus esculentus*)	80	8	120	90–150	100	good	semi-cross (insects)
Onion (*Allium cepa*)	4	20	200	420[3]	600	poor	cross (insects)
Peppers (*Capsicum annuum*)	5	0.6	150	45–60	100	good	semi-cross (insects)
Pumpkin (*Cucurbita spp.*)	150	3	200	60–120	600	good	cross (insects)
Radish (*Raphanus sativus*)	12	8	60	420[3]	600	good	cross (insects)
Tomato (*Lycopersicon esculentum*)	3	0.05	1000	45–60	5	good	self

Note:
(1) the figures for 1000 grain weight and multiplication factor are general estimates only. Different varieties of beans, broad beans and many others differ considerably in seed size (1000 grain weight), and multiplication factors largely depend on yields, which in turn depend on the potential of the area and the cultivation practices. Small seeded varieties generally have a higher multiplication factor than large seeded types of the same crop.
(2) Seeding rate and multiplication factor of rice depend largely on the cultivation method: direct sowing or transplanting. Biennials take more than a year from seed to seed, except in areas with bimodal rainfall patterns where seed may be produced within 12 months.
(3) Seed viability when stored at appropriate temperature and humidity (see Table 7.1).

PART E

APPENDIXES

Appendix 1. Glossary and Abbreviations

The definitions given in this glossary refer to the terms commonly used in the seed sector. The same terms may have slightly different meanings in another context.

accession   an entry in a genebank; a sample of seed (or plants) representing a particular variety or population.

adaptability   the capability (of a variety) to perform well under given agro-ecological conditions (see wide adaptation, yield stability).

advanced material   plants in a breeding programme that show a real deviation from the ancestral materials, but that are not yet a variety.

allele   a physical form of a gene, e.g. part of a chromosome coding, for example the flower colours white, red, yellow, etc.

apomixis   a type of vegetative reproduction through seed-like structures (e.g. in certain grass species).

biodiversity   the variability among living organisms from all sources, including terrestrial, marine and other aquatic ecosystems and the ecological complexes of which they are part; this includes diversity within species, between species and of ecosystems (Agenda 21, CBD, Rio de Janeiro, 1992).

blending   the intentional mixing of seed lots.

bolting   the emergence of a seed stalk in biennial crops.

buffer crop   crop, planted around a seed field to protect the latter from foreign pollen.

centre of diversity   geographical area where a wide genetic diversity is found for particular crops and related species.

certification   the assurance of varietal identity and varietal purity in seed production through generation control, inspection, and labelling.

certified seed   a seed class in a certification scheme, produced from Foundation or Registered (AOSCA-system) or Basic (OECD-system) seed, which is sold to farmers for crop production.

CGIAR   Consultative Group on International Agricultural Research, being the umbrella organization for a number of International Agricultural Research Centres (IARCs) throughout the world.

chromosome   (in higher plants) a thread-like structure in the nucleus of cells, composed of DNA molecules, that contains the genes. Chromosomes are species specific in number and type.

clone   genetically homogeneous population of a vegetatively propagated crop.

co-evolution	the evolution of plants, diseases, pests and weeds in interaction with each other and with human activities.
composite variety	a population of a cross-fertilizing crop resulting from a mixture of selected components (lines or populations).
conditioning	all treatments of seeds from harvesting to planting, e.g. drying, storage, dressing, packaging.
contract grower	farmer producing seed for a seed organization on contract.
cross-fertilizer	a crop where under normal conditions seeds are produced by intercrossing between plants within a population.
cultivar	see 'variety'.
degeneration	the loss of seed quality through subsequent generations. This can be due to the build-up of seed-transmitted diseases or through genetic changes due to mutations, introgression or admixture that results in unwanted characteristics such as small grain, lack of uniform product, etc. Degeneration can be avoided through improved seed production technology and maintenance breeding.
domestication	the genetic adaptation of wild plants to the cultivation environment through selection.
dormancy	the condition that a viable seed does not germinate when supplied with those factors normally considered adequate for germination.
double cross-hybrid	hybrid, resulting from a cross between two single cross-hybrids.
D.U.S.	Distinctness, Uniformity (homogeneity), Stability. The requirements for a variety in a certification system and used for plant variety protection.
emasculation	removal of potentially effective male organs (e.g. in hybrid production).
emergence	the development of full-grown seedlings in the field (see also germination, viability, vigour).
FAO	Food and Agriculture Organisation of the United Nations.
farm-saved seed	seed sown at the same farm where it was harvested.
farmers' rights	the rights that local farmers and communities can claim over genetic resources in landraces.
field inspection	inspection of a seed field to check on isolation, seed crop management, presence of diseases, varietal purity, etc.
formal seed supply system	seed supply through an organized chain of events by specialized breeders, seed producers, marketing agents, and including certification.

274

gene	a unit of inheritance; a region in a chromosome that encodes for a particular trait, e.g. flower colour.
genotype	the entire genetic constitution of an organism
genebank	a genetic resources centre where genotypes and populations are stored as seeds, pollen or plants.
generation system	a method in formal seed production whereby a seed lot can be traced back to a particular lot of breeders' (pre-basic) seed. This forms the basis of a certification system.
genetic diversity	the genetic variation within and between populations of species.
genetic drift	random genetic changes in a population due to small population size.
genetic erosion	the global loss of genetic diversity, or the loss of genetic diversity in crops in a particular farming system.
genetic resource	synonym for germplasm with emphasis on its actual or potential value as a resource.
germination	the resumption of development and growth of a fully developed embryo, starting with the uptake of water and ending when photosynthesis starts. (In seed testing: the percentage of seeds of a seed lot developing within a given time into normal seedlings.) (see also, emergence, viability, vigour).
germplasm	any living material; can be used for propagation and breeding purposes, emphasis on its genetic content.
Green Revolution	the large-scale introduction of new farming technologies in the (sub-) tropics in the 1960s and 70s, notably short straw rice and wheat in combination with fertilizers and pesticides.
heterogeneous	mixed, variable. When used in genetics, referring to a population consisting of a mixture of genotypes.
heterosis	the superiority in performance of a hybrid compared to the mid-parent value of both parents. It is often used for the cases in which the hybrid is superior to both parents.
heterozygous	the combination of two (or more) different alleles on the same locus in one genotype.
hybrid	general: the first generation progeny of a cross between two different parents. In seed production: a variety, of which the seed is produced by controlled crossing of two different parents (see also: single cross, three-way cross, double cross, top cross).
hybridization	the crossing of two different plants with the aim to create diversity (in a breeding programme), or to produce hybrid seed.
IARC	International Agricultural Research Centre.

275

inbred line	genetically (nearly) homogeneous and (nearly) homozygous population, used for hybrid seed production.
inbreeding	self-fertilization, or mating of individuals that are more closely related genetically than individuals meeting at random. In cross-fertilizing crops this may lead to inbreeding depression.
inbreeding depression	the loss of vigour and fitness as a result of self-fertilization in species that are normally cross-fertilized.
in situ conservation	preservation of genetic resources in the area where they naturally occur, i.e. in nature or in farmers' fields.
integrated seed supply system	combination of different aspects of the formal and the local seed supply system, aimed at improvement of performance of both systems.
introgression	the integration of foreign alleles in a population through cross-fertilization (with other variety plants or related weeds)
ISTA	International Seed Testing Association.
landrace	a variety developed by farmers in particular agro-ecological and socio-economic conditions, usually a complex, heterogeneous population.
line	progeny of a single (self-fertilized) plant.
mating system	the common way of combining male and female gametes in a species: self-fertilization, (semi) cross-fertilization.
mini-kit	small quantity of different components of a new technology (e.g. seeds, fertilizer, pesticides, information leaflets), distributed to increase the adoption rate.
multi-line variety	mixture of similar but not genetically identical pure lines of a self-fertilized crop, selected for improved performance.
multiplication factor	general: the number of seeds produced from one parent seed. In seed production: net seed yield per hectare (i.e. after seed cleaning and quality control), divided by the seed rate.
mutation	a spontaneous variation of the genetic material of a cell.
OECD	Organization for Economic Co-operation and Development.
off-type	a plant differing from the variety in morphological or other trait, e.g. as a result of segregation, mutation, cross-fertilization or mechanical mixing. Such plants are either removed (rogued) or nurtured (as possible improvement of the variety).

on-farm research	strategy of formal agricultural research, whereby farmers are involved in problem definition, setting of research priorities, testing and selection of technologies.
on-farm trials	the testing component of on-farm research.
open-pollinated variety (OP)	variety multiplied through random fertilization, i.e. opposite to hybrid (commonly used for cross-fertilized species only)
parastatal	company, operating under public administration rules and regulations, of which business losses are replenished by government.
phytosanitary control	measures to avoid the introduction of foreign plant diseases.
plant breeders' rights	the legal right of the originator of a modern crop variety within a plant variety protection system.
plant population	the number of plants per unit area, expressed e.g. as number per hectare. See also: population.
plant variety protection	legal system of granting exclusive rights over varieties to the originator (breeder or discoverer).
population	a group of individuals that share a common gene-pool and have the potential to interbreed.
post-control	a final quality check on seed lots, used as an internal quality control of the seed quality control organization.
product life cycle	in seed supply: the period in which a variety is adopted by farmers, popular with many farmers, and gradually replaced by competing (newer) varieties.
product mix	in seed supply: the array of crops and varieties offered by a seed supplier.
pure-line variety	genetically homogeneous and homozygous variety of a self-fertilizing crop.
roguing	removal of individual plants from a seed field, because they are off-type or diseased.
seed	generative or vegetative part of a plant that is used as propagation material.
seed aid programme	emergency assistance programme, supplying seed to disaster-affected farmers.
seed board	regional or national committee to direct seed production and supply initiatives.
seed chain	the successive operations leading to seed supply, i.e. breeding, seed production, conditioning, marketing and quality control.
seed class	denomination of a generation within a certification scheme.
seed cleaning	removal of unwanted inert matter and seeds from a seed stock.
seed dressing	improvement of seed performance through chemical or other treatment of the seed coat.

seed policy	statement by a government to guide the development of seed supply.
self-fertilized crop	a crop where under normal conditions seeds are produced as a result of self-fertilization in at least 95 per cent of the cases.
seed security programme	activity designed to avoid the loss seed by large numbers of farmers.
seggregation	the appearance of different plants in an offspring due to the separation of different alleles in the mother plant. Segregation frequently occurs during the first generations after a cross, and less frequently later.
semi-cross-fertilized crop	a crop where the majority of seeds are produced as a result of self-fertilization but varying percentages of outcrossing may occur up to 50 per cent depending on external conditions.
single cross hybrid	cross between two inbred lines.
synthetic variety	a variety produced by crossing a number of genotypes selected for their good combining ability.
three-way cross hybrid	hybrid resulting from a cross between an inbred and a single cross hybrid.
top cross hybrid	hybrid resulting from a cross between an inbred line and a population or open pollinated variety.
true to type	a plant which conforms with the description of the variety it belongs to (opposite to off-type).
UPOV	International Union for the Protection of New Varieties of Plants.
varietal hybrid	hybrid resulting from a cross between two open-pollinated populations (population hybrid).
variety	a plant grouping within a botanical taxon, which can be defined by the expression of characteristics resulting from a given genotype or combination of genotypes, and sufficiently homogeneous to be distinguished from other such groupings by the expression of at least one characteristic. Synonymous with cultivar.
variety maintenance	the conservation of the important features of a variety through continuous selection.
variety release	the official approval of a variety for multiplication and distribution.
viability	the ability of a seed to germinate and to produce a normal seedling when conditions are good (see germination, vigour).
vigour	the properties which determine the potential for rapid, uniform emergence and development of normal seedlings under a wide range of field conditions (see germination, viability).
V.C.U.	Value for Cultivation and Use; i.e. the combined values of a variety, established by (field) trials.

weed   a plant which grows where people do not want it to grow; it can be a wild species or a cultivated species growing in another crop.

wide adaptation   the ability of a variety to perform well under a variety of agro-ecological conditions.

yield stability   low variation in crop yield over time in a particular place, even where seasons differ considerably (see adaptation, wide adaptation).

Appendix 2: Crop names

English	Botanical	French	Spanish	Portuguese
Amaranth	*Amaranthus* spp.	amaranthe	amaranto, bledo	caruru vermelho
Barley	*Hordeum vulgare* L.	orge	cevada	cebada
Bean (common bean)	*Phaseolus vulgaris* L.	haricot commun	judia comun	feijado
Cabbage	*Brassica oleracea* L.	chou vert	col-berza	couve
Carrot	*Daucus carota* L.	carotte	zanahoria	cenoura
Cassava	*Manihot esculenta* Crantz	manioc	yuca brava, manioca	mandioca
Chick pea	*Cicer arietinum* L.	pois chiche	garbanzo	grao de bica
Chilli/sweet pepper	*Capsicum annuum*	poivron, piment	pimiento	pimento
Cowpea	*Vigna unguiculata* (L.)Walp.	niebe, dolique	frijol arroz	feijado chicote
Cucumber	*Cucumis sativus* L.	concombre/cornichon	pepino	pepino
Eggplant	*Solanum melongena* L.	aubergine	berengena	beringela
Finger millet	*Eleusine coracana* (L.)Gaertn.	eleusine	mijo coracana	capim naxenim
Foxtail millet	*Setaria italica* (L.)Beauv.	millet des oiseaux	panizo comun	milho painco
Green gram	*Vigna radiata* (L.) Wilczek	amberique	judia mung	ervilha Jerusalem
Groundnut	*Arachis hypogaea* L.	arachide	cacahuete/mani	amendoim
Lentil	*Lens culinaris* Med.	lentille	lenteja	lentilha
Lettuce	*Lactuca sativa* L.	laitue	lechuga	alface
Maize	*Zea Mays* L.	maïs	maiz	milho
Melon	*Cucurbita melo* L.	melon	melon	melao
Oats	*Avena sativa* L.	avoine	avena	aveia
Okra	*Abelmoschus esculentus* (L.)Moench.	gombo/okra	gombo/ocra	quiabo
Onion	*Allium cepa* L.	oignon	cebolla	cebola
Pea	*Pisum sativum* L.	pois	guisante/arveja	ervilha
Pearl millet	*Pennisetum americanum* (L.)	mil a chandelle	mijo perla/junco	milheto massango
Pigeon pea	*Cajanus cajan* (L.)Millsp.	ambrevade	guando	guandu
Potato	*Solanum tuberosum* L.	pomme de terre	patata/papa	batata
Pumpkin	*Cucurbita* spp.	potiron	abobora menina	calabaza amarilla
Radish	*Raphanus sativus* L.	radis	rabano	rabanete
Rice	*Oryza sativa* L./ *O.glaberrima* Steud.	riz	arroz	arroz
Rye	*Secale cereale* L.	seigle	centeno	centeio
Sesame/benniseed	*Sesamum indicum* L.	sesame	sesamo	gergelim
Sorghum	*Sorghum bicolor* (L.)Moench	sorgho, gros mil	sorgo	sorgo
Soybean/soya bean	*Glycine max* (L.)Merr.	soja	soja	soja
Sunflower	*Helianthus annuus* L.	tournesol	girasol	girassol
Tomato	*Lycopersicon esculentum* Mill.	tomate	tomate	tomate
Wheat	*Triticum aestivum* L.	blé/froment	trigo candeal	trigo mole
Yam	*Dioscorea* spp.	igname	name/yame	inhame

Appendix 3: References

Benzing, A. 1989. Andean potato peasants are seed bankers. *ILEIA News-letter* Vol. 5 No.4 12–13.

Berg, T., 1992. Indigenous knowledge and plant breeding in Tigray, Ethiopia. *Forum for Development Studies* No. 1: 13–22.

Bertuso, A., and R. Salazar, (in prep). Community genebanks and farmer breeding: the experience of CONSERVE in the Philippines. In: Almekinders, C, and W. de Boef (eds.), *Towards a synthesis between crop conservation and development*. IT Publications, London.

Brush, S.B., H.J. Carney and Z. Huaman, 1981. Dynamics of Andean potato agriculture. *Economic Botany* 35: 70–88.

Burfisher, M.E. and N.R Horenstein, 1985. Sex roles in the Nigerian Tiv farm household. Women's roles and gender differences in development: cases for partners No. 2. Connecticut, Kumarian Press.

Byerlee, D., 1994. Modern varieties, productivity, and sustainability: recent experience and emerging challenges. CIMMYT, Mexico.

Cárdenas, H. and P. Alvarez, 1996. Producción de semilla de frijol y maíz de buena calidad en la zone de Puriscal. In: P. Alvarez and H. Cárdenas, *Biodiversidad vegetal y manejo local*. Seminario. IDEAS, San José.

Chrispeels, M.J. and D.E. Sadava, 1994. *Plants, genes and agriculture*. Jones & Barlett Publishers, England.

Commutec, 1996. Socio-economic and technical factors determining community biodiversity and management. SADC/GTZ project. Promotion of small-scale seed production by self-help groups, Harare.

Cromwell, E. (ed), 1990. Diffusion mechanisms in small farm communities: lessons from Asia, Africa and Latin America. Network Paper 21, ODI, London.

DANAGRO, 1988. SADC regional seed production and supply project. Main report (Vol. A) and country reports (Vol. 2A-2J). DANAGRO Advisor, Glostrup, Denmark.

Dennis, J.V., 1987. Farmer management of rice diversity in northern Thailand. PhD Thesis, Cornell University. Ithaca, New York.

Ertug Firat, A., and A. Tan, 1997. *In situ* conservation of genetic diversity in Turkey. In: N. Maxted, B.V. Ford-Lloyd and J.G. Hawkes (eds.), *Plant genetic conservation: The* in situ *approach*. London, Chapman & Hall, p. 254–262.

FAO, 1996. *Global Plan of Action for the conservation and sustainable utilisation of plant genetic resources for food and agriculture*. FAO, Rome.

Feldstein H.S., and S.V. Poats, 1990. *Working together. Gender analysis in agriculture. Volume 1: Case studies*. Kumarian Press, West Hartford.

Ferguson, A.E., and R.M. Mkandawire, 1993. Common beans and farmer-managed diversity; regional variations in Malawi. *Culture and Agriculture* Vol. 45/46: 14–17.

Greve, J.E. van S., 1983. Safe storage for small quantities of seed. *Harvest* 9: 5–10.

Grisley, W., and M. Shamambo, 1993. An analysis of the adoption and diffusion of Carioca beans in Zambia, resulting from experimental distribution of seed. *Experimental Agriculture* 29: 379–386.

Gusson, M.F., 1998. Melhoramiento de milho em experiência participativa. In: A.Campolina Soares, Altari Toledo Machado, Breno de Mello Silva and Jean Marc von der Weid (ed.) Milho crioulo conservaçâo e uso da biodiversidade, AS-PTA, Rio de Janeiro.

Johannessen, C.L., 1982. Domestication process of maize continues in Guatemala. *Economic Botany* 36: 86–99.

Kornegay, J., J. Alonso Beltran, and J. Ashby, 1996. Farmers' selections within segregating populations of common bean in Columbia. In: P. Eyzaguirre and M. Iwanaga, *Participatory plant breeding.* Proceedings of a workshop on participatory plant breeding, 26-29 July 1995, The Netherlands. IPGRI, Rome.

Magnifico, F., 1996. Community-based resource management: CONSERVE (Philippines) experiences. In: L. Sperling and M. Loevinsohn (eds.) *Using Diversity. Enhancing and maintaining genetic resources on-farm.* Proceedings of a workshop held on 19-21 June, New Delhi, India. New Delhi, IDRC, p. 289–301.

Martin G., and M.W. Adams, 1987. Landraces of *Phaseolus vulgaris* (Fabaceae) in Northern Malawi. II. Generation and maintenance of variability. *Economic Botany* 41: 204–215.

Maurya, D.M., 1989. The innovative approach of Indian farmers. In: R. Chambers, A. Pacey and L.A. Thrupp (eds.). *Farmer first: Farmer innovation and agricultural research.* London. IT Publications, p. 9–13.

Mheen Sluijer J., 1996. Towards household seed security. SADC/GTZ project Promotion of small-scale seed production by self-help groups, Harare.

ODI, 1996. Seed provision during and after emergencies. Good Practice Review 4. ODI Seeds and biodiversity programme, London.

Oosterhout, S. van, 1996. Coping strategies of smallholder farmers with adverse weather conditions regarding seed deployment of small grain crops during the 1994/1995 cropping season in Zimbabwe. Volumes 1 to 3, SADC/GTZ, Harare, Zimbabwe.

Osborne, T., 1995. *Participatory agricultural extension: experiences from West Africa.* Gatekeeper Series 48, IIED, London, pp. 19.

RAFI, 1986. *The community seed bank kit.* RAFI, Canada.

Scheidegger, U., G. Prain, F. Ezeta, C. Vittorelli, 1989. *Linking formal R & D to indigenous systems: a user oriented potato seed programme for Peru.* Agricultural Administration Network Paper 10, ODI, London, pp 20.

Sperling, L., and J.A. Ashby, 1997. Participatory plant breeding: emerging models and future development. In: R. Tripp (ed.), *New seeds and old laws. Regulatory reform and the diversification of national seed systems.* IT Publications, London.

Sperling, L., and U. Scheidegger, 1995. Participatory selection of beans in Rwanda: results, methods and institutional issues. Gatekeeper Series No. 51. IIED, London.

Toledo Machado, A., (in prep.). The collaborative development of stress tolerant maize varieties in Brazil. In: Almekinders, C. and W. de Boef (eds.), *Encouraging diversity: Crop development and conservation in plant genetic resources.* IT Publications, London.

UNEP, 1992. Convention on Biological Diversity. June 1992. United Nations Environment Program, Nairobi.

Wierema, H., C. Almekinders, L. Keune and R. Vermeer, 1993. *La producción campesina en Centroamerica. Los sistemas locales de semilla.* IVO, Tilburg, The Netherlands.

Worede, M., 1997. Ethiopian *in situ* conservation. In: N. Maxted, B.V. Ford-Lloyd and J.G. Hawkes (eds.), *Plant genetic conservation. The in situ approach.* London, Chapman & Hall, p. 290–301.

Worede, M., and H. Mekbib, 1993. Linking genetic resource conservation for farmers in Ethiopia. In: de Boef, W., K. Amanor, K. Wellard and A. Bebbington (ed.), *Cultivating knowledge: Genetic diversity, farmer experimentation and crop research*, IT Publications, London, p 78–84.

Wright, M., T. Donaldson, E. Cromwell and J. New, 1994. The retention and care of seeds by small farmers. NRI, Chatham.

Zimmerer. K., 1989. Seeds of peasant subsistence: agrarian structure, crop ecology and Quechua agriculture in reference to the loss of biological diversity in the southern Peruvian Andes. PhD Thesis, Univ. Of California, Berkeley.

Appendix 4: Recommended reading

Genetic Resources and Biodiversity

Cooper, D., R. Vellve and H. Hobbelink, 1992. *Growing diversity: genetic resources and local food security.* IT Publications, London.

Ford-Lloyd, B. and M. Jackson, 1986. *Plant Genetic Resources. An introduction to their conservation and use.* Edward Arnold, London.

Frankel, O.H. and J.G. Hawkes, 1975. *Crop genetic resources for today and tomorrow.* Cambridge University Press.

Frankel, O.H. and M.E. Soulé, 1981. *Conservation and evolution.* Cambridge University Press.

Harlan, J.R., J.M.J. de Wet and A.B.L. Stemler, 1976. *Origins of African plant domestication.* Mouton, The Hague.

Harlan, J.R., 1992. *Crops and Man,* 2nd ed. American Society of Agronomy, Madison.

Heide, W.M. van der, and R. Tripp and W.S. de Boef (ed.), 1996. *Local crop development: an annotated bibliography.* IPGRI, Rome/CPRO-DLO(CGN), Wageningen.

Heywood, V.H., and R.T. Watson, 1995. Global Diversity Assessment, UNEP, Cambridge University Press.

IPGRI, 1993. *Diversity for development: the strategy of the International Plant Genetic Resource Institute.* IPGRI, Rome.

Kloppenburg, J.R., 1988. *First the seed: the political economy of plant biotechnology, 1492–2000.* Cambridge University Press.

Querol, D., 1992. *Genetic resources. Our forgotten treasure. Technical and socio-economic aspects.* Third World Network, Malaysia. 252 pp.

Zohani, D. and M. Hopf, 1993. *Domestication of plants in the old world, 2nd ed.* Clarendon Press, Oxford.

Local Seed Production Systems

Almekinders, C.J.M., N.P. Louwaars and G.H. de Bruijn, 1994. Local seed systems and their importance for an improved seed supply in developing countries. *Euphytica* No. 78: 207–216.

Boef, W. de, K. Amanor, K. Wellard and A. Bebbington (ed.), 1993. *Cultivating Knowledge. Genetic diversity, farmer experimentation and crop research.* IT Publications, London.

Cherfas, J., M. and J. Fanton, 1996. *Seed savers' handbook.* Grover Books, U.K.

Clawson, D.L., 1985. Harvest security and intraspecific diversity in traditional tropical agriculture. *Economic Botany* No. 39: 56–67.

Cromwell, E. (ed.), 1990. Diffusion mechanisms in small farm communities: lessons from Asia, Africa and Latin America. Network paper 21, ODI, London.

Hernández, E., 1985. Maize and man in the Southwest. *Economic Botany* 39 Vol. No. 4: 416–430.

Merrick, L., 1989. Crop genetic diversity and its conservation in traditional agroecosystems. In: M.A. Altieri and S.B. Hecht (ed.), *Agroecology and small farm development,* pp. 3–11. CRC Press, Boca Raton, U.S.A.

Improvement of Local Seed Systems

Cromwell, E., S. Wiggins and Sandra Wentzel. 1993. *Sowing beyond the state. NGOs and Seed Supply in developing countries.* ODI, London.

Cromwell, E., E. Friss Hansen and M. Turner, 1992. The seed sector in developing countries; a framework for performance analysis. ODI Working paper 65. London.

Louwaars, N.P. and G.A.M. van Marrewijk. *Seed supply Systems in Developing Countries.* CTA, Wageningen.

Osborne, T., 1990. Multi-institutional approaches to participatory technology development: a case study from Senegal. ODI Network paper 13, London.

Sperling, L., M.E. Loevinsohn and B. Ntambovura, 1993. Rethinking the farmer's role in plant breeding: local bean experts and on-station selection in Rwanda. *Expl. Agr.* No. 29: 509–519.

Participatory Approaches, Gender Analysis

Ashby, J., 1990. *Evaluating technology with farmers. A handbook.* IPRA, CIAT, Cali, Colombia.

Chambers, R. A. Pacey and L.A. Trupp (ed.), 1989. *Farmer First. Farmer innovations and agricultural research.* IT Publications, London.

Feldstein, H.S., and S.V. Poats, 1990. *Working Together. Gender analysis in agriculture. Volume 1: case studies.* Kumarian Press, West Hartford

Pretty, J.N., I. Guijt, J. Thompson and Ian Scoones, 1995. *A trainer's guide for participatory learning and action* and: *A user's guide for participatory learning and action.* IIED Participatory Methodology Series, London.

Reijntjes, C., B. Harverkort and A. Waters-Bayer. 1992. *Farming for the future. An introduction to low-external-input and sustainable agriculture,* Macmillan, London.

Experiences with Participatory Crop Improvement and Seed Production

Ashby, J.A., C. A. Quiros and Y.M. Rivers. 1989. Farmer participation in technology development: work with crop varieties. In: R. Chambers, A. Pacey and L.A. Thrupp (ed.), *Farmer first. Farmer innovation and agricultural research,* pp. 115–132. IT Publications, London.

Cromwell, E., S. Wiggins and S. Wentzel, 1993. *Sowing beyond the state: NGOs and seed supply in developing countries.* ODI, London.

Eyzaguirre, P. and M. Iwanaga (ed.), 1996. Participatory plant breeding. Proceedings of a workshop on participatory breeding, 26–29 July 1995, Wageningen, The Netherlands. IPGRI, Rome.

Joshi, A. and J.R. Witcombe, 1996. Farmer participatory crop improvement. II. Participatory varietal selection, a case study in India. *Expl. Agric.* No. 32: 461–477.

Prain, G., F. Uribe and U. Scheidegger, 1991. Small farmers in agricultural research: farmers' participation in potato germplasm evaluation. In: Haverkort, B., J. van der Kamp and A. Waters-Bayer (ed.), *Joining farmers' experiments. Experiences in participatory technology development.* IT Publications, London.

Sperling, L. and J.A. Ashby, 1997. Participatory plant breeding: emerging models and future development. In: Tripp, R. (ed.), *New seed and old laws. Regulatory reform and the diversification of national seed systems.* IT Publications, London.

Sperling, L., U. Scheidegger and R. Buruchara, 1993. Designing bean seed systems for smallholders. *ILEIA Newsletter* 9(1): 24–25.

Sthapit, B.R., K.D. Joshi and J.R. Witcombe, 1996. Farmer participatory crop improvement. III. Participatory plant breeding, a case study for rice in Nepal. *Expl. Agric.* 32: 479–496.

Thiele, G., G. Gardner, R. Torrez and J. Gabriel, 1997. Farmer involvement in selecting new varieties: potatoes in Bolivia. *Expl. Agric.* 33: 275–290.

Whitcombe, J.R., A. Joshi, K.D. Joshi and B.R. Sthapit, 1996. Farmer participatory crop improvement. I. Varietal selection and breeding methods and their impact on biodiversity. *Expl. Agric.* 32: 445–460.

General Formal Seed Production Technology

Agrawal, R.L., 1980. *Seed Technology.* Oxford and IBH Publ. Co. New Delhi.

Copeland, L.O. and M.B. McDonald, 1985. *Principles of seed science and technology.* Burgess Publ. Co. Minneapolis.

CTA and IAC, 1985. *Proceedings of the seminar on seed production, Yaounde, Cameroun, 21–25 October, 1985.* 2 Volumes. CTA, Wageningen.

Doerfler, T., 1976. *Seed production guide for the tropics.* Colombo, Dept. of Agric., Peradeniya, Sri Lanka.

Gastel, A.J.G. van, M.A. Pagnotta, E. Porceddu (ed.) (1996). *Seed science and technology.* ICARDA, Aleppo, Syria.

Hebblethwaite, P.D. (ed.), 1991. *Seed production.* Butterworths, London.

Kelly, A.F., 1989. *Seed planning and policy for agricultural production: the roles of government and private enterprise in supply and distribution.* Belhaven, London.

Marshall, C. and J. Grace, 1992. *Fruit and seed production: aspects of development, environmental physiology and ecology.* Cambridge University Press.

Srivastava, J.P. and L.T. Simarski, 1986. *Seed production technology.* ICARDA, Aleppo, Syria.

Stoll, G. (1986). *Natural crop protection based on local farm resources in the tropics and subtropics.* Margraf, Weickersheim, Germany.

Thomson, J.R., 1979. *An introduction to seed technology.* Leonard Hill, Glasgow.

Wellving, A.H.A., 1984. *Seed production handbook of Zambia.* Dept.of Agriculture, Lusaka.

Processing and Storage

Chin, H.F. and E.H. Roberts, 1980. *Recalcitrant crop seeds.* Tropical Press, Kuala Lumpur.

Delhove, G.E. and W.L. Philpott, 1983. *World list of seed processing equipment.* FAO, Rome.

FAO, 1981. *Cereals and grain legume seed processing.* FAO, Rome.

Jeffs, K.A., 1988. *Seed treatment.* The British Crop Protection Council, London.
Justice, O.L. and J.H. Bass, 1979. *Principles and practices of seed storage.* Castle House, Tunbridge Wells, UK.
Linnett, B., 1986. *Seed processing in Australia.* Queensland Dept of Primary Industries information series Q186008. QDPI, Brisbane.
Roberts, E.H., 1972. *Viability of seeds.* Chapman & Hall, London.

Seed Quality Control

AOSA (1993). *Rules for testing seeds. Journal of Seed Technology* 16(3).
Bekendam, J. and R. Grob, 1979. *Handbook for seedling evaluation.* ISTA, Zürich.
Burg, W.J. van der, J. Bekendam, A. van Geffen and M. Heuver, 1983. Project seed laboratory 2000–5000. *Seed Sci. & Technol* 11: 157–227.
Felfoldi, E.M., 1983. *Handbook of pure seed definitions.* ISTA, Zürich.
Hampton, J.G. and D.M. TeKrony (ed.).(1995). *ISTA handbook of vigour test methods.* ISTA, Zürich, Switzerland.
ISTA, 1982. *Survey of equipment and supplies for seed testing.* ISTA, Zürich.
ISTA 1999. *International rules for seed testing.* ISTA, Zürich, Switzerland.
Mathur, S.B. and J. Jorgensen, 1992. *Seed pathology.* CTA, Ede.
Mayer, A.M. and A. Poljakov-Mayber, 1989. *The germination of seeds,* 4th ed. Pergamon, Oxford.
Neergaard, P., 1977. *Seed pathology.* Macmillan Press, London.
OECD, *Seed schemes for field crops, oil crops, herbage and pasture crops* (different volumes), OECD, Paris.
Richardson, M.J., 1979. *An annotated list of seed-borne diseases.* ISTA, Zürich.

Crop Technical Information

Chopra, K.R., 1982. *Technical guidelines to sorghum and millet seed production.* FAO, Rome.
Feistritzer, W.P. (ed.), 1982. *Technical guidelines for maize seed technology.* FAO, Rome.
Feistritzer, W.P. and A. Fenwick Kelly, 1987. *Hybrid seed production of selected oil and vegetable crops.* FAO, Rome.
George, R.A.T., 1985. *Vegetable seed production.* Longman, London.
Kelly, A. F., 1988. *Seed production of agricultural crops.* Longman, London.
Maessen, L.J.G. van der and Sadikin Somaatmadja (ed.), 1989. *Plant resources in South-east Asia (PROSEA) 1; Pulses.* Pudoc, Wageningen.
National Research Council Board on Science and Technology for International Development, 1996. *Lost crops of Africa.* Volume 1. National Academy Press, Washington
Purseglove, J.W., 1977a. *Tropical crops, dicotyledons.* Longman, London.
Purseglove, J.W., 1977b. *Tropical crops, monocotyledons.* Longman, London.
Salumke, D.K., B.B. Desai and N.R. Bhat, 1987. *Vegetable and flower seed production.* Agricole Publ. New Delhi.

Siemonsma, J.S. and Kasem Piluek, 1993. *Plant resources in South-east Asia (PROSEA) 8: Vegetables.* Pudoc, Wageningen.

Skerman, P.J., D.G. Cameron and F. Riveros, 1988. *Tropical forage legumes.* FAO, Rome.

Skerman, P.J. and F. Riveros, 1990. *Tropical grasses.* FAO, Rome.

Willan, L.A., 1985. *A guide to forest seed handling with special reference to the tropics.* FAO, Rome.

Policy and Legislation

Bombin-Bombin, L.M., 1980. *Seed legislation.* FAO, Rome.

Douglas, J.E., 1980. *Successful seed programmes. A planning and management guide.* Westview Press, Boulder, Colorado

Grall, J. and B.R. Levi, 1985. *La guerre des semences. Quelles moissons, quelles sociétés.* Artyheme Fayard, Paris.

Jaffee, S. and J. Srivastava, 1992. *Seed system development: the appropriate roles of private and public sectors.* World Bank Discusson Papers 167, IBRD, Washington.

Macmullen, N., 1987. *Seeds and world agricultural progress.* Nat. Planning Assoc., Washington.

Mooney, P.R., 1980. *Seeds of the earth.* Inter Pares, Ottawa.

Pearse, A., 1980. *Seeds of plenty, seeds of want: social and economic implications of the Green Revolution.* Clarendon Press, Oxford.

Rosell, C.H. and A. Fenwick Kelly (ed.), 1983. *Seed campaigns. Guidelines for promoting the use of quality seeds in developing countries.* FAO, Rome.

Tripp, R. (ed.), 1997. *New seeds and old laws. Regulatory reform and the diversification of national seed systems.* ODI/IT Publications, London.

Appendix 5. Addresses

Relevant organisations dealing with seeds

AOSA: Association of Official Seed Analysts. c/o USDA National Seed Storage Laboratory, Colorado State University, Fort Collins, CO 80523, USA. www.zianet.com/AOSA

AOSCA: Association of Official Seed Certifying Agencies in North America and New Zealand, POBox 81152, Lincoln NE 685011152, USA.fax +1 402 476 6547; e-mail: assoc@navix.net

ASSINSEL, Association Internationale des Selectionneurs Professionnels pour la Protection des Obtentions Végétales, Chemin de Reposoir 5–7, 1260 Nyon, Switzerland.

CGIAR: Consultative Group on International Agricultural Research, 1818 H Street NW, Washington DC 20433, USA.

EU: (Commission of the) European Union, Rue de la Loi 100, 1049 Brussels, Belgium.

FAO: Food and Agriculture Organisation of the United Nations. Seeds and Plant Genetic Resources Service, Via delle Terme di Caracalla, 00100 Rome, Italy. www.fao.org

FIS: Federation Internationale du Commerce des Semences, Chemin de Reposoir, 1260 Nyon, Switzerland.

ISTA: International Seed Testing Association. Reckenholz, Postfach 412, 8046 Zürich, Switzerland.fax +41 1 371 34 27. e-mail: istach@iprolink.ch

OECD: Organisation for Economic Co-operation and Development, Directorate for Food, Agriculture and Fisheries, 2 Rue André Pascal, 75775 Paris Cedex 16, France.

UPOV: Union Internationale pour la Protection des Obtentions Végétales, 34 Chemin des Colombettes, 1211 Geneve, Switzerland. www.upov.org

World Bank, Agricultural and Rural Development Department, 1818 H Street NW, Washington DC 20433, USA. www.worldbank.org

Institutes with seed-related research (CGIAR)

CIAT: International Centre for Tropical Agriculture, Apdo Aereo 6713, Cali, Colombia.

CIFOR: Center for International Forestry Research, Bogor, Indonesia.

CIMMYT, International Centre for Maize and Wheat Improvement, Londres 40, Apdo Postal 6–641 Mexico DF.

CIP: International Potato Center, Apdo 5969, Lima, Peru.

CPRO-DLO, POBox 16,6700 AA Wageningen, The Netherlands, fax: +31 317 418094, e-mail: wsc@cpro.dlo.nl

IPGRI: International Plant Genetic Resources Institute, Via delle Sette Chiese 142, 00142 Rome, Italy.

ICARDA: International Center for Agricultural Research in the Dry Areas. POBox 5466, Aleppo, Syria.

ICRAF: International Centre for Research in Agro-Forestry. POBox 30677, Nairobi, Kenya.

ICRISAT: International Centre for Research in the Semi-Arid Tropics, Patancheru PO, Andhra Pradesh 502324, India.

IITA: International Institute for Tropical Agriculture, PMB 5320, Ibadan, Nigeria.
ILRI: International Livestock Research Institute: POBox 5689, Addis Ababa, Ethiopia and POBox 30709, Nairobi, Kenya.
INIBAP: International Network for the Improvement of Banana and Plantain. Agropolis-Montpellier, Bat.7 – Boulevard de la Lironde, 34980 Montferrier sur Lez, France.
IRRI: International Rice Research Institute: POBox 933, Manilla, Philippines.
ISNAR: International Service for National Agricultural Research. Postbus 93375, 2509 AJ, The Hague, The Netherlands.
WARDA: West African Rice Development Association. 01 BP 2551, Bouake, Ivory Coast.

Institutes with regular international training courses in seed technology
CIRAD: Centre de Coopération Internationale en Recherche Agronomique pour le Développement. BP 5035, 34032 Montpellier Cedex, France.
CPRO-DLO/CGN, POBox 16,6700 AA Wageningen, The Netherlands, fax: +31 317 418094, e-mail: dus-testing@cpro.dlo.nl
DSE/ZEL: Wielingerstrasse 52, 8133 Feldafing, Germany.
Danish Government Institute for Seed Pathology in Developing Countries, Ryvangs Alle 78, DK-2900 Hellerup-Copenhagen, Denmark.
Edinburgh School of Agriculture, West Mains Road, Edinburgh EH9 3JG, Scotland, UK.
IAC: International Agricultural Centre, POBox 88, 6700 AB Wageningen, The Netherlands.
ICARDA (in conjunction with University of Jordan, Amman): POBox 5466, Aleppo, Syria.
Instituto Agronomico Mediterraneo de Zaragoza. Apartado 202, 50080, Zaragoza, Spain.
Maize Research Institute Zemun Polje, POBox 89, 11081, Zemun, Serbia.
Massey University, Seed Technology Centre, Palmerston North, New Zealand.
Mississippi State University, POBox 5267, MS 39762, USA.
UPLB: University of the Philippines in Los Banos, Seed Technology Programme, POBox 430, Los Banos, College, Laguna, Philippines.
Svalof AB, S-28800 Svalov, Sweden
Zamorano University, POBox 93, Tegucigalpa, Honduras

Other institutions/projects with seed-related and development-oriented research and activities
Agromisa, P.O. Box 41, 6700 AA Wageningen, e-mail: agromisa@nld.toolnet.org For questions on small-scale agriculture, publication of Agrodok series
CTA, Centre Technique de Coopération Agricole et Rurale, P.O. Box 380, 6700 AJ Wageningen, The Netherlands. e-mail: cta@cta.nl Information for agriculture and rural development Newsletter: *Spore.*
ECHO, 17430 Durrance Road, North Fort Myers, FL 33917 2239, USA, fax: +1–941-543–5317, e-mail: echo@xc.org, www..xc.org/echo For questions on small scale agriculture, publications, etc.

ETC/ILEIA: P.O. Box 64, 3830 AB Leusden, The Netherlands. ILEIA Newsletter. e-mail: ileia@ileia.nl www.etcint.nl
IIED: International Institute for Environment and Development, 3 Endsleigh Street, London, WC1H ODD, UK. e-mail: iiedinfo@gn.apc.org
Grain: Girona 25 pral, E-8010 Barcelona, Spain, Newsletter: Seedling, www.grain.org
NRI: Natural Resources Institute. Central Avenue, Chatham Maritime, Kent, ME4 4TB, U.K.
ODI: Overseas Development Institute. Portland House, Stag Place, London SW1E 5DP, U.K. ODI Network Papers, *Agrinews.*
RAFI: Rural Advanced Foundation International, 110 Osborne St. Suite 202, Winnipeg, MB R3L 1Y5, e-mail: rafi@rafi.org www.rafi.org
On-Farm Productivity Enhancement Program OFPEP (in Senegal, Uganda, Kenya, Ethiopia). c/o Bird Building, Western Carolina University, Cullowhee, North Carolina, USA. Newsletter: *Of soils and seeds*, e-mail: pvouc@wcu.edu
On the Small Scale Seed Production by Self-Help groups, SADC/GTZ project (Harare). Newsletter: *Local seed system news.* e-mail:onsssp@samara.co.zw, www.zimbabwe.net/sadc-fanr/intro.htm

Additional relevant information sites on the World Wide Web
www.undp.org
www.seedtest.org   documents on seed testing, including ISTA.
www.cgiar.org   provides access to the wwweb pages of the CG institutions like IPGRI, CIMMYT, CIAT, etc. These pages have information on their activities and research results, names and addresses to contact.
www.oneworld.org   a site with access to information and newsletters of English organisations such as IT Publishers, ODI, etc.
www.idrc.ca   a valuable site of IDRC-Canada with information on activities and publications. Various publications can be read and printed entirely.
www.cpro.dlo.nl   site of Centre for Genetic Resources, The Netherlands, with information on literature references, publications and activities related with agrobiodiversity, conservation *in situ.*
www.biodiv.org   site of the CBD (Convention on Biodiversity).

www.ingramcontent.com/pod-product-compliance
Lightning Source LLC
Chambersburg PA
CBHW052010030426

42334CB00029BA/3153